About Island Press

Since 1984, the nonprofit organization Island Press has been stimulating, shaping, and communicating ideas that are essential for solving environmental problems worldwide. With more than 1,000 titles in print and some 30 new releases each year, we are the nation's leading publisher on environmental issues. We identify innovative thinkers and emerging trends in the environmental field. We work with world-renowned experts and authors to develop cross-disciplinary solutions to environmental challenges.

Island Press designs and executes educational campaigns in conjunction with our authors to communicate their critical messages in print, in person, and online using the latest technologies, innovative programs, and the media. Our goal is to reach targeted audiences—scientists, policymakers, environmental advocates, urban planners, the media, and concerned citizens—with information that can be used to create the framework for long-term ecological health and human well-being.

Island Press gratefully acknowledges major support of our work by The Agua Fund, The Andrew W. Mellon Foundation, The Bobolink Foundation, The Curtis and Edith Munson Foundation, Forrest C. and Frances H. Lattner Foundation, The JPB Foundation, The Kresge Foundation, The Oram Foundation, Inc., The Overbrook Foundation, The S.D. Bechtel, Jr. Foundation, The Summit Charitable Foundation, Inc., and many other generous supporters.

The opinions expressed in this book are those of the author(s) and do not necessarily reflect the views of our supporters.

RESTORING NEIGHBORHOOD STREAMS

Restoring Neighborhood Streams

Planning, Design, and Construction

Ann L. Riley

ISLANDPRESS

Washington | Covelo | London

Island Press is a trademark of The Center for Resource Economics.

Library of Congress Control Number: 2015952388

✪ Printed on recycled, acid-free paper

Manufactured in the United States of America
10 9 8 7 6 5 4 3 2 1

Keywords: Island Press, restoration, urban stream restoration, neighborhoods

This book is dedicated to Lisa Owens Viani,
neighborhood activist and environmentalist

CONTENTS

In concept, urban stream restoration is not new, it is not radical, and it is not something that many would consider as being undesirable or unrealistic. Why, then, is it that such little progress is being made on restoring urban streams when fundamentally we all understand the multiple benefits that can be realized from reduced flood damages, improved water quality, enhanced ecosystems, and a resource to which people can connect physically, socially, and, in some cases, spiritually? Setting aside momentarily the willingness to invest in stream restoration, the primary limiting factors are the tools, examples, and confidence needed to move into areas of uncharted planning and design that may look risky and maybe even daunting to designers and decision makers. Since the 1980s, groups have been urging the creation of such tools and training, with limited success. During this time, the Association of State Floodplain Managers (ASFPM) has strongly advocated for floodplains, both in terms of reducing flood risk and in terms of enhancing and restoring natural functions too often degraded using traditional methods of design. The ASFPM has also partnered with Ann Riley on multiple ventures and has come to understand her vision, knowledge, and passion as being integrally entwined with urban stream restoration. We are pleased and grateful about this book, her latest contribution to educate and lead by example on how to approach urban stream restoration. This book helps fill the gaps for the planning and design tools we all need.

Doug Plasencia, PE, CFM
President, ASFPM Foundation

My first acknowledgment is to the Association of State Floodplain Managers Foundation, which funded a large portion of the costs to produce the graphics for this book. I am also grateful to the association for its effectiveness in advancing river and floodplain management in the United States. All the graphics in this book, unless otherwise noted, were prepared by graphic artist Lisa Krieshok, who did a masterful job of interpreting my clumsy sketches, scribbles, and construction drawings. Technical advisors were Pete Alexander and Joe Sullivan, fish biologists with the East Bay Regional Park District; Leslie Ferguson and Janet O'Hara, water quality engineers, and Kevin Lunde, environmental scientist, with the San Francisco Bay Regional Water Quality Control Board; Roger Leventhal, Farwest Restoration Engineering; Lee Liang, hydrologist with the Santa Clara Valley Water District; Dennis O'Connor, principal of Habitat Concepts, Portland, Oregon; and Liza Prunuske, principal of Prunuske-Chatham Inc.

The following people contributed to the stream restoration cases: Danny Akagi, City of Berkeley Public Works Department; Mitch Avalon, past deputy director of Contra Costa County Public Works Department; Joshua Bradt, past executive director of the Urban Creeks Council and current codirector of the San Francisco Bay California Urban Streams Partnership; Leslie Estes, program manager for watersheds and storm drainage, City of Oakland Engineering Department; Arlene Fong, Friends of Glen Echo Creek; Drew Goetting, principal of Restoration Design Group; Norman La Force, past mayor of the City of El Cerrito; Skip Lisle, president of Beaver Deceiver International; Melanie Mintz, community development director for the City of El Cerrito; Yvetteh Ortiz, Public Works Department, City of El Cerrito; Lisa Owens Viani, founder of Friends of Baxter Creek; Heidi

Perryman, founder of Worth a Dam, City of Martinez; Lynne Scarpa, environmental manager of the Stormwater Program, City of Richmond; Carol Schemmerling, cofounder of the Urban Creeks Council; Mike Vukman, past restoration director of the Urban Creeks Council and current codirector of the California Urban Streams Partnership; and Valerie Winemiller, Steering Committee of the Piedmont Avenue Improvement League.

Chapter 1

Is The Restoration of Urban Streams Possible?

"It's just an urban stream," said the engineering consultant, responding to my request to vegetate the channel rather than line it with plastic geogrid. We are taught that restoring ecologically functioning urban streams and rivers is not possible, based on the belief that urban watersheds are too degraded and their landscapes too altered to support naturally functioning systems. Restoring urban streams and rivers is also not possible, we are told, because it is prohibitively expensive to practice ecological restoration in a setting where land is expensive and other land uses are valued more highly than streams. Restoration is not possible, the argument continues, because the public will not accept the flood and erosion hazards associated with uncontrolled dynamic natural streams in the interiors of cities.

Engineers, landscape architects, and planners are taught this framework in college and graduate school. The instruction includes urban stormwater literature from which students get the impression that after about a 10 to 15 percent increase in imperviousness from urbanization, it is likely that we reach a point of no return for salvaging a stable, ecologically functioning stream. Some researchers make definitive conclusions; one is that in watersheds where impervious cover exceeds 60 to 70 percent, it is not going to be impossible to restore streams (Clayton 2000).

A preponderance of urban stormwater literature shows how land use changes affect urban flood hydrographs: streams flow faster after rain, channels enlarge and erode, and large floods happen more frequently. We witness how urbanization fills in headwater streams, encases channels in concrete, puts channels and drainages in culverts, and permanently alters the drainage network of the natural stream system. Channels can incise and widen sometimes as much as eight times their original size (Hammer 1972). Eroding channels have simplified environments so that biological diversity is reduced: aquatic insects, fish, reptiles, and amphibians may barely survive and probably not thrive. Riparian plant communities are destroyed and degraded and are invaded by exotic species that can crowd out natives. The wildlife dependent on these plant communities disappears. These observations are indisputable (Bernhardt and Palmer 2007).

The rational individual therefore sees little potential in restoring urban streams, other than fostering public education and improving the urban quality of life. Even some of the most open-minded, supportive professionals who appreciate the urban environmental movement arising out of the 1980s and 1990s do not have expectations beyond increasing the aesthetic values of urban streams.

In 1982, I decided to address the issue of whether the restoration of urban streams and rivers, including the most degraded, is possible. It became clear to me that the only way to test this hypothesis was to believe that it was possible and set out to try. This book is written to record the results of this thirty-year experiment. Obviously, this effort necessarily required anyone entering this ambitious and time-consuming experiment to have a bias that it *is* possible to restore urban streams. I have attempted to honestly, and in some cases brutally, report project failures, ridiculous naiveté, and how better restoration practices evolved out of making mistakes. I cover the history of how the earliest smaller-neighborhood-scale projects came about and how they became the experimental settings for developing restoration design protocols. The projects described here are in urban watersheds ranging from 0.2 square mile to 16.7 square miles with limited project lengths of 200 to 750 feet. These smaller-scale projects subsequently set the stage for later large-scale projects measured in miles.

The question of whether restoring urban streams and rivers is possible must address three basic challenges. First, given the degraded urban watershed conditions and land use constraints inherent in the city, is it physically feasible to return a degraded stream to an ecologically functioning and dynamic state? Natural streams are inherently dynamic environments and require erosion, deposition, moving and adjusting plan forms, and flooding to be truly living streams. Are living streams and these urban conditions mutually exclusive concepts? The second challenge is whether it is financially feasible or reasonable to attempt to re-establish this type of dynamic ecosystem in a city. The third challenge is to ask whether enough public support can be developed to enable the sometimes inconvenient land use changes that may be necessary to allow for a functioning, live stream.

The urban stream and river case studies described in this book are organized around these questions: Did the project result in a geomorphically and biologically functioning stream? Could the project substitute for an engineered channel to provide a solution to flooding and excessive erosion? Were the identified benefits of the project achieved at a reasonable cost? How were land uses in conflict with achieving a functioning ecosystem system resolved? The case studies address these ecological, economic, and social issues and fairly answer the question of whether urban stream restoration is possible.

For the case studies to be credible, they need to be located in highly modified watersheds, represent lower- or middle-income communities with limited economic resources, and occur where the classic conflicts between city life, land use, and erosion and flood hazards exist. Without these factors, the information gained from these case studies cannot be transferable to other heavily urbanized environments. Most of the case studies in this book are therefore located in working-class, low-income, or poverty neighborhoods. They are in areas where restoration concepts conflict with housing, streets and parking, and recreational needs and in areas where the safety of children must be a concern, such as school grounds and parks. All are located where there have been flooding hazards.

How Urban Streams Differ from Streams in Other Settings

Memorable field trips taken to the rural California Sacramento and Tuolumne Rivers and their tributaries overwhelmed me with the cumulative challenges of rivers riprapped (rocked), straightened, logged, dredged, dammed, diverted to agricultural fields, and constrained with berms and levees composed of toxic mining tailings. This widened perspective made restoring an urban stream in a city look welcomingly feasible. Streams constantly adjust and attempt to recover from human modifications, no matter what the setting is. The streams work to rebalance the sediment loads, discharges, shapes, and slopes, and sometimes the plant community recovers over time on its own. At times, stream "restoration" is a matter of a stream directing its own recovery. In other circumstances, humans intervene in an effort to hasten or redirect the recovery process. In some cases, restoration occurs because of the intervention of animals such as beavers. Environmental management professionals need to keep in mind that the channel evolution and recovery processes we observe in more rural environments are also present in urban environments. The variables making up stream dynamics—including topography, rainfall, discharges, sediment loads and sizes, vegetation, and valley and channel slopes—are present, even for the streams encased in concrete. That means that the restoration practitioner in an urban environment may be able to reasonably describe how a stream may respond to changes in discharges, sediment supplies, channel slopes, and vegetation removal. Diagnostic assessments can be carried out to identify watershed and stream system problems such as risk of flood damages or excessive erosion, and they help remedy the causes of the imbalances, not just put a Band-Aid on them.

Many professionals promote stereotypes or unquestioned assumptions about urban streams or rivers. One of the most notorious symbols of the highly degraded urban river in the United States is the Los Angeles River, encased in concrete since the 1930s. A number of Los Angeles flood control engineers state that this river cannot feasibly recover any natural functions because the upstream debris basins and concrete channels prevent river sediment transport. Without a sediment supply, the river cannot conceivably begin to express natural channel forms or river dynamics. A field trip to the Los Angeles River not only reveals a channel filled with a wide range of sediment sizes ranging from sands to cobbles and boulders, but also a sediment supply sufficient for creating channel complexity and allowing willow thickets to re-establish (fig. 1.1). I was also led by local officials to the rectangular concrete channel Coyote Creek (fig. 1.2) and the trapezoidal channel San Jose Creek (fig. 1.3), both tributaries to the San Gabriel River, the other degraded urban river draining the Los Angeles area. We observed channels with substantial sediment supplies that have been transported and deposited to form a single-thread meandering channel with a floodplain that supports riparian vegetation. This situation is occurring to the consternation of flood control officials who view this re-creation of natural depositional forms within the flood control channel as an unfortunate maintenance problem. Many other flood con-

FIGURE 1.1 The North Fork Los Angeles River is re-forming in a dirt-bottom flood control channel at Glendale Narrows, Los Angeles.

FIGURE 1.2 Coyote Creek in the San Gabriel watershed, Los Angeles County, is recovering ecological functions within a rectangular, concrete flood control channel through sediment transport, deposition, and recovery of riparian vegetation.

trol engineers administering to other similar flood control channels sympathize with this commonly occurring "problem."

Water quality conditions can also take an ironic twist in urban areas. The Los Angeles San Gabriel Watershed Council monitoring program shows high levels of coliform bacteria in the more natural upper watershed areas used for recreation than in downstream reaches in urbanized Los Angeles (Belden 2008). Similar findings from a water quality agency study on Wildcat Creek in Richmond, California, indicate pockets of pollution located in the protected open-space regional and local parklands. That pollution can be attributed to the large number of dog walkers who do not clean up the excrement of their pets (Surface Water Ambient Monitoring Program 2008).

Many highly urbanized streams still have access to sediment supplies from less developed or underdeveloped steep headwater areas and the bed and banks of the channels. Some receive regular supplies from naturally unstable hillsides, landslides, and fault zones or alluvial fans. Although urban streams in highly developed environments can be expected to have lower sediment supplies than they did as natural systems, it is not good practice to universally assume that urban streams

FIGURE 1.3 San Jose Creek in the San Gabriel watershed, Los Angeles County, is recovering a channel, floodplain, and riparian corridor in a trapezoidal flood control channel.

are cut off from all supplies or that sediment supply is automatically a limiting factor precluding the return of some of the natural functions of both erosion and deposition and sediment transport.

In a case like the Los Angeles River, increasing the sediment transport and deposition functions may well be one of the important strategies for increasing the functioning of a highly urbanized river corridor. The Los Angeles River contains 8 miles of a dirt-bottom river through the Glendale Narrows section of the river. This reach flows downstream between the city of Burbank, upstream of Griffith Park, and Taylor Yard, a defunct railroad maintenance yard located just above downtown Los Angeles. The Taylor Yard reach near Glendale and the 101 Freeway in figure 1.4 is also referred to as Frog Town by locals because this dirt-bottom channel supports vegetation and a braided channel type with the physical complexity sufficient to support an amazing number of insects, amphibians, and bird species. No flood problems have occurred along these dirt-bottom reaches, and they provide a model for removing the concrete channel inverts along the other reaches of the river.

Although many planning efforts focused on identifying restoration options for

FIGURE 1.4 The Los Angeles River at "Frog Town," Los Angeles, supports birds and aquatic habitat for amphibians in a dirt-bottom flood control channel at the 101 Freeway. *Credit: Los Angeles Department of Water and Power.*

the Los Angeles River have struggled with design strategies for bringing life back to the river, one obvious model for restoration already exists in the Glendale Narrows section. Short floodwalls can be added to the higher terrace to contain the expected higher water-surface elevations for the largest flood flows, which are caused by the changes in the river cross sections as it fills with sediment and vegetative growth. Concrete reaches can emulate the dirt-bottom sections with removal of the concrete invert and use of grade controls to support the concrete sides while allowing the river to have a functioning ecosystem in the bottom portion of the channelized system (Friends of the Los Angeles River 1995). My favorite location along the Los Angeles River is shown in figure 1.5 near Frog Town, where the dirt-bottom channel transitions again to a concrete bottom and the river is carving out a single-thread "active channel" transporting water and sediment through the concrete.

The Los Angeles River was historically represented by a number of channel and wetland types, including a single-thread and braided channel with freshwater floodplain marshes and tidal marshes. In some areas, the river would meander up to 7 miles. The river will most likely never return to its original historic form, although the information on the historic landscape is being used to set new restoration objectives in opportunity areas (Stein et al. 2007). The current constraints on the Los Angeles River do not mean, however, that it cannot function as a different type of river within its confined state and provide ecological "services" as well as improved aesthetics, as illustrated in figure 1.6 (Garrett 1993).

By the 1980s, tertiary treated reclaimed water turned the Los Angeles River into a perennial river. In 2013, after years of advocacy by the Friends of the Los Angeles River, the River Project, Heal the Bay, the City of Los Angeles, and others, a stunning recognition for the environmental potential for the river was achieved: The US Army Corps of Engineers completed seven years of studies and planning and released the Los Angeles River Ecosystem Feasibility Study, which adopted a number of restoration projects as feasible. The numerous alternatives in the study build off the soft-bottom reaches for 11 miles from the confluence of Arroyo Seco with the river upstream of the Griffith Park area. The study identified opportunistic projects for land acquisition along the river to widen the corridor and restore ecological function. Now, restoring the Los Angeles River is no longer a joke but a city imperative (MacAdams 2013). A coordinated effort involving citizen scientists and professionally trained biologists working together to inventory

FIGURE 1.5 Nature bats last, forming a meander through the concrete bottom of the Los Angeles flood control channel.

Figure 1.6 The Los Angeles Flood Control Project features an extensive riparian corridor at Riverside Drive near Figeroa Street north of Interstate 5 and the 101 Freeway.

the wildlife in and near the river and write papers on local biodiversity is currently under way. Through this effort, a biological inventory of the river published by the Natural History Museum of Los Angeles County in 1993 (Garrett 1993) is being updated. We now know that the Griffith Park area supports mule deer, bobcat, raccoon, striped skunk, coyote, and mountain lion. Insects dependent on river environments—including the Andrena bee, which is willow dependent—are also present. Native fish—including the Santa Ana sucker, speckled dace, arroyo chub, and rainbow trout—occupy the river's headwater streams. The lower Los Angeles River supports one of the largest stopover, wintering, and breeding areas for shorebirds in coastal Southern California. In addition, historical ecological information is being gathered to help with functional restoration on the river, with The Nature Conservancy taking the lead with other partners to start a project to enhance habitat along 2.5 miles of the river. The biodiversity potential for the river is no longer marginalized by professional scientists (Council for Watershed Health 2014). The jury is still out on whether future projects will achieve functional ecosystem restoration that may be perceived by some Los Angeles interests to compete with economic, development, and recreational objectives, but the joke (and at one time serious proposal) about turning the river into a new freeway is over.

How Urban Streams Evolve

Another widely believed misperception is that once an urban stream is impacted by urban development, the natural processes are so radically compromised that the typical channel evolution that streams go through to recover from imbalances in less developed environments rarely happens. The well-known river scientist Luna Leopold addressed this issue by recording the ability of small urbanizing river basins to evolve and adjust over time to urbanization toward a new equilibrium. Shortly before his death in 2006, Leopold reflected on his data collected over a period of forty years in the urbanizing watershed of the Watts Branch in Maryland and compared his observations to those made by others of urbanizing basins (Leopold, Huppman, and Miller 2005). Leopold's findings were later confirmed by Anne Chin (Chin 2006), who reviewed more than a hundred case studies documenting stream adjustments to urbanization over five decades.

Leopold describes what he calls the urbanization cycle of urban streams and stresses that the time periods described below are approximate and vary among cities and regions. The first stage, with a typical duration of ten years, occurs when the watershed and stream channel experience initial development with housing, roads, and then some sewer construction. At this stage, flood discharges increase, but the urbanizing stream is still resilient. The second stage he describes generally occurs within the next ten years. It is one in which many construction sites are cleared and erosion leads to large sediment loads delivered to stream channels. As a result, the typical stream channel builds massive point bars and deposits sediment over its banks, and stream channels then narrow. A third stage is characterized by development occurring over another twenty years, often involving straightening and burying streams in pipes and lining them in concrete and or rock. Toward the end of this cycle, there is a shortage of building sites. The older neighborhoods are well vegetated and are hydrologically stabilized, but the stable state reflects the large area of roofs and pavement. This period is characterized by more and larger peak flows and a decrease in sediment load, typically producing wider channels. The fourth stage occupies another decade or so when new building nearly ceases and the neighborhoods and their landscapes mature. Channels are wide and migrate by erosion where banks are not stabilized by trees or revetments.

Leopold, Huppman, and Miller (2005) describe the fifth and last stage of urbanization as a period in which the public wishes stream channels to be revitalized and made more natural. As they describe: "Some reaches are exhumed at great expense. Water temperatures are cooler because of shading by trees. Some fish return."

A review of urban stream research from around the world indicates that urban streams can and do reach equilibrium after the development phase is largely completed and after impervious surfaces and the artificial drainage network (usually denser) have stabilized so that the runoff and sediment regimes reach a new equilibrium. Many urban streams undergoing study are still in adjustment phases and can go through periods of stability until new watershed changes impact that stability. Researchers agree that the core principles for managing urban streams are to understand the evolution of urban streams and identify the phase they are in. Understanding this evolutionary process also helps identify the context for any restoration effort, which cannot necessarily assume returning to a preurbanization state (Chin 2006).

On the basis of her review of the available research, Anne Chin concludes:

> Considerable interest in the published literature has focused on the question of whether urban streams can truly adjust to changed hydrologic conditions and reach new stability regimes. The range of research and investigations has demonstrated that this is indeed possible for most rivers, given the degrees of freedom with which to adjust, albeit over longer periods in some. (Chin 2006)

As in any environmental setting with land use changes, urban streams evolve at times according to the "classic" incised channel evolution model. In this model, when urban streams are straightened, they react by "headcutting" or eroding down the bed in an up-channel direction to flatten slopes; experience bank failure and widen; and rebuild floodplains and stable active channels within a wider cross section (Schumm, Harvey, and Watson 1984; Simon 1989). Sometimes urban streams follow other evolutions in process and form, including recovery through a combination of channel widening, meandering, and incision or through formation of steps and pools.

The operative principle is that it is hard to generalize about our urban watersheds. We must develop an awareness of which stream processes are acting on the channel and the inherent differences for recovery rates. It is not constructive to lump many or all highly urbanized watersheds into the "once it's fully developed, it's all over for recovery" category.

The urban stream and river restoration movement is still in its infancy, with the first thirty-plus years of records coming from pioneering regions such as the San Francisco Bay area, where the case studies in this book are located. It is hoped that researchers can build on the records we have kept on these projects and that even greater understanding of the "ecological, economic, and social services" of urban stream and river restoration projects can be developed.

Clarifying Different Perspectives on Restoration

Part of the task in evaluating whether restoration is possible is to bring clarity to what the term *restoration* means. Because it means different things to different professionals, it creates confusion in the literature. Because of the importance of developing clarity on this term, chapter 2 is devoted to establishing the definition of restoration used to evaluate the case studies presented here.

The first principle I apply is that restoration should not be confused with beautification or aesthetic "improvements." Restoration, in fact, may not be aesthetic to some people and can entail fallen trees, willow thickets, downed wood in streams, and other features that may cause concern to a neighbor looking at the stream in a local park. For purposes of evaluating the benefits of restoration, we need to determine if the attributes of a "natural" stream—such as channels, pools, riffles, vegetation, and depositional features, including floodplains—are present. We also need to know whether these forms contribute to a range of natural processes and dynamics that streams should exhibit, such as the transport of sediment, erosion, deposition, floodplain flooding, and nutrient exchange. If the stream is not dynamic and changing, it is not a living stream. Stream restoration, whether in a rural or urban setting, should not be rock gardening in which we constrain the stream dynamics with boulders and other hardscapes and then add landscaping with potted plants from the nursery.

The related principle is to improve the functioning of the stream and floodplain so that it can contribute to higher water quality and habitat conditions sufficient for supporting some aquatic and terrestrial habitat. Stream landscape forms and dynamics can improve water quality in much the same manner as a water treatment plant can (Riley 2009). In some situations in urban and rural areas alike, returning functions may need more watershed-scale management of polluted runoff from farms, forest harvest sites, or urban streets to achieve the desired water quality and habitat objectives. In urban environments, scientists recommend combining stormwater management and habitat restoration to achieve a better functioning habitat. Stormwater management projects may still be in progress, and the management of sewage and stormwater discharges may not be complete, but that need not prevent stream and floodplain restoration projects from moving forward. Doing so can subsequently create the public will to better address these related needs, and the projects can eventually complement one another.

Applying Local Examples to Other Restorations

In 1994, Thomas "Tip" O'Neill, who served in Congress from 1953 to 1986 and was a very productive Speaker of the House for ten years, wrote a book called *All*

Politics Is Local (O'Neill and Hymel 1994). The same can be said about stream restoration. Typically, the drivers for the projects are the needs to solve local problems such as bank erosion, flood risks, and property damages. Sometimes people organize themselves around saving green spaces, creating parks, or protecting "charismatic" flora and fauna such as steelhead fish or raptors. O'Neill's book contains pointers for aspiring politicians that any restoration consultant, environmental activist, or local official would be wise to heed. Chapter headings such as "People Like to Be Asked," "Don't Create Opponents for Yourself," "Persistence Pays," "You Can Teach Old Dogs New Tricks," and "Compromise Is the Art of Politics" reveal advice for anyone interested in accomplishing an environmental restoration project. Certainly the cases described here indicate that successful political and community organizing processes are common to each, and it is hard to imagine that these principles are not relevant in many areas of the world.

The cases in this book are located in California. What will readers from another place—Massachusetts, Florida, Europe, or Australia, for instance—learn of value to their different environments? First, they will learn about the history and evolution of the stream restoration practice, which affects all locales. Chapter 2 will help readers anywhere grapple with what is a good working definition of restoration. The description of the various schools of restoration applied to the design of the projects will be of use to people no matter what the project is and who the readers are. Projects often involve application of hydraulic engineering, fluvial geomorphology, and native plant and fish biology. Readers who are seasoned practitioners, teachers, or students looking for exposure to the field will find that the restoration cases will provide a thoughtful review of how these fields may be applied in complementary ways to inform restoration design. Chapter 3 tells the story of six urban stream restoration projects that have inspired a number of additional related projects. These cases should provide universally interesting information on how the projects were conceived and organized and how the science applied to restoration design evolved over time. Finally, chapter 4 discusses factors common to many restoration projects, including the range of costs, monitoring and assessments, and maintenance issues.

Case Study Geography and Demographics

This book focuses on six case studies of urban stream restoration projects and related spin-off projects located in the San Francisco Bay region of north coastal California illustrated in figure 1.7. In several instances, the six projects motivated more related projects in their watersheds, and those are also described here. The watershed areas represent a subregion of the bay that receives approximately 23 to 25 inches of average annual rainfall precipitation. All the creeks and river projects described here flow into the San Francisco Bay and are a product of the Mediterranean climate in which the precipitation occurs mostly between October and May. Light snow only occasionally occurs on the highest elevations in hills more than 1,000 feet above sea level, so the cases described here are usually not affected by snowfall runoff. The population of the nine-county area considered the Bay Area contains about 7 million people. All the cases are located in densely populated urban cities. Oakland, Berkeley, Albany, El Cerrito, San Pablo, and Richmond all merge into one another's boundaries and represent a continuous and dense urban corridor with no open-space buffers between cities. These cities are located approximately 6 miles across the bay from the San Francisco. The Martinez case is located in the San Francisco delta subregion.

Innovations in urban stream restoration projects in the East San Francisco Bay began in 1982–1983 on Strawberry Creek in Berkeley, an early daylighting project, and in 1984 on Glen Echo Creek in Oakland, an early effort to substitute a flood control project with a restoration approach. Both creeks are located in residential neighborhoods. The Village Creek and Codornices Creek restoration projects, located in Albany and Berkeley, respectively, were implemented between 1999 and 2010. These two creek restoration projects share a similar location and history as part of a housing redevelopment project on property owned by the University of California. This book describes the Village Creek case rather than the Codornices Creek case, which is a larger, regional-scale project that does not meet the definition of a neighborhood-scale project, the focus of this book. The1995 Blackberry Creek project, located in Berkeley, was sponsored by the local parent-teacher association for a schoolyard setting. Figure 1.8 shows the projects in these watersheds, all of which are located in Alameda County.

Demographically, these projects are located in communities of poor to moderate-income neighborhoods in a range of residential and commercial settings. The Baxter Creek projects span moderate-income neighborhoods to poverty and working-class-income areas. The Strawberry Creek neighborhood, a crime hot spot before the creek restoration and park development project, evolved into a solidly middle-class neighborhood after the project was completed. The Glen Echo Creek project in Oakland represents a neighborhood surrounded by large, busy commercial thoroughfares, including car sales lots and an overhead freeway, but it attains a middle-class status because of historic architecture and the buffering effect of the Glen Echo Creek greenbelt. The Blackberry Creek and Baxter Creek daylighting projects are located in middle-class neighborhoods in Berkeley and El Cerrito, respectively. The setting for the Village Creek project is a low-income housing development for university students on the border of Berkeley and Albany. The downtown Martinez business district provides the setting for the

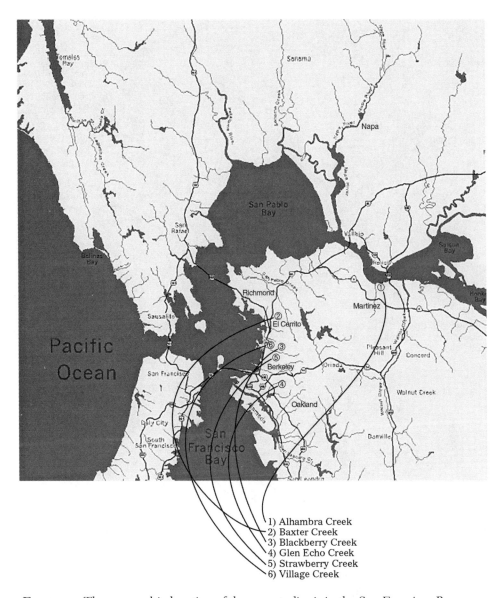

1) Alhambra Creek
2) Baxter Creek
3) Blackberry Creek
4) Glen Echo Creek
5) Strawberry Creek
6) Village Creek

FIGURE 1.7 The geographic location of the case studies is in the San Francisco Bay region of California. *Credit: Lisa Kreishok.*

Alhambra Creek restoration project and represents a small-town working-class to middle-class environment.

At 78 square miles and with a population of 400,740, Oakland, the site of the Glen Echo Creek project, is the third largest city in the Bay Area. It historically has been an important Bay Area transportation hub, and the Port of Oakland is

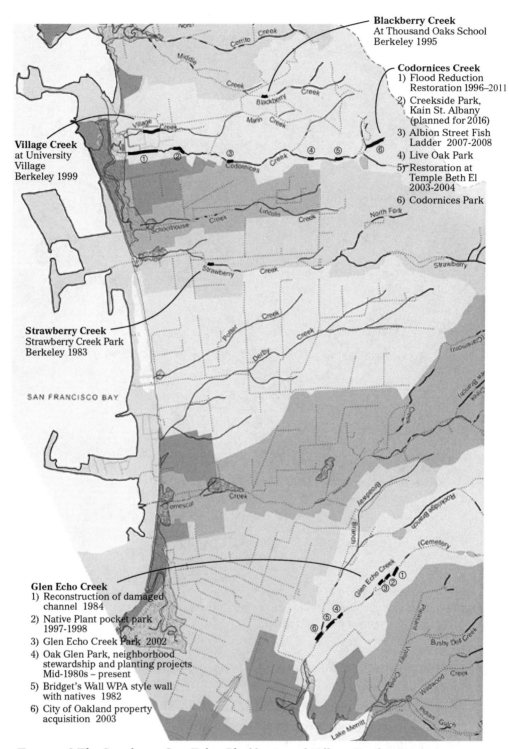

Blackberry Creek
At Thousand Oaks School
Berkeley 1995

Codornices Creek
1) Flood Reduction
   Restoration 1996–2011
2) Creekside Park,
   Kain St. Albany
   (planned for 2016)
3) Albion Street Fish
   Ladder 2007-2008
4) Live Oak Park
5) Restoration at
   Temple Beth El
   2003-2004
6) Codornices Park

Village Creek
at University
Village
Berkeley 1999

Strawberry Creek
Strawberry Creek Park
Berkeley 1983

SAN FRANCISCO BAY

Glen Echo Creek
1) Reconstruction of damaged
   channel 1984
2) Native Plant pocket park
   1997-1998
3) Glen Echo Creek Park 2002
4) Oak Glen Park, neighborhood
   stewardship and planting projects
   Mid-1980s – present
5) Bridget's Wall WPA style wall
   with natives 1982
6) City of Oakland property
   acquisition 2003

FIGURE 1.8 The Strawberry, Gen Echo, Blackberry, and Village Creeks projects are located in Oakland, Berkeley, and Albany in Alameda County. *Credit: Lisa Kreishok.*

one of the largest ports in the West, competing with Long Beach, California, and Seattle, Washington. Oakland is divided into approximately fifty neighborhoods. The city's population is almost evenly divided between Americans who are white, African, Asian, and Hispanic. As is the case with all these East Bay cities, the high-income households live in the hills to the east (the hills are about 1,000 to 1,700 feet at the highest elevation), whereas the lower-income households live in the flatlands adjacent to the bay. African American settlement increased greatly during World War II in the flatlands of all these cities, but particularly in Oakland and Richmond. The per capita annual income in Oakland is a relatively low $22,000, and 19 percent of its residents live below the poverty line, with 28 percent of those under eighteen years old living in poverty conditions. The crime rate continues to be an issue for this city.

Neighboring Berkeley is contained in an area of 17.7 square miles and has a population of 112,000 divided into about twenty-five neighborhoods. It is best known as a university town, but has industry located in the flatlands. The Blackberry Creek project is in a high-end north Berkeley neighborhood. This project also serves a lower-income bracket of families located down the hill in west Berkeley whose children attend the school on the project site. The Village Creek project forms the floodplain environment along with Codornices Creek where the University of California, Berkeley's family student housing complex is located. This west Berkeley–west Albany community represents a 1940s movement to integrate white, African American, and Asian American populations into housing and work environments associated with the World War II shipyard industry and military facilities. The Strawberry Creek daylighting project converted a crime-ridden, weed-strewn railroad right-of-way into a park, which helped accomplish a major neighborhood upgrade. At $38,896, the per capita income in Berkeley is a much higher than that of Oakland; still, a relatively high 18 percent of the population lives below the poverty level. About 55 percent of the Berkeley population is white, 19 percent Asian, 9.7 African American, and 10 percent Hispanic. The population is composed of an international mix of many cultures.

Continuing north on the map in Figure 1.8 is the small city of Albany with a population of 18,500 within 5.5 square miles. It has a very high cost of living and a per capita income of $39,400. It is 51 percent white, 10 percent Hispanic, and 3.5 percent African American. Like Oakland and Berkeley, the demographics in Albany change significantly between the hills and flatlands populations. In 2007, the lower watershed was approximately 45 percent white, 31 percent African American, and 13 percent Hispanic, whereas the upper watershed areas was 84 percent white, 4 to 5 percent Hispanic, and 2 percent African American. The upper watershed median household income was $111,000, whereas the lower watershed household income was $42,850 (Watershed Project and Codornices

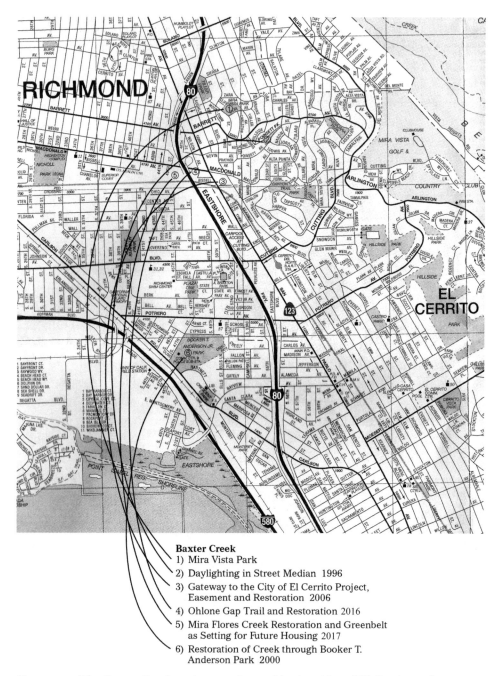

Baxter Creek
1) Mira Vista Park
2) Daylighting in Street Median 1996
3) Gateway to the City of El Cerrito Project,
   Easement and Restoration 2006
4) Ohlone Gap Trail and Restoration 2016
5) Mira Flores Creek Restoration and Greenbelt
   as Setting for Future Housing 2017
6) Restoration of Creek through Booker T.
   Anderson Park 2000

FIGURE 1.9 The Baxter Creek projects are located in the cities of El Cerrito and Richmond, Contra Costa, Co. *Credit: Lisa Kreishok.*

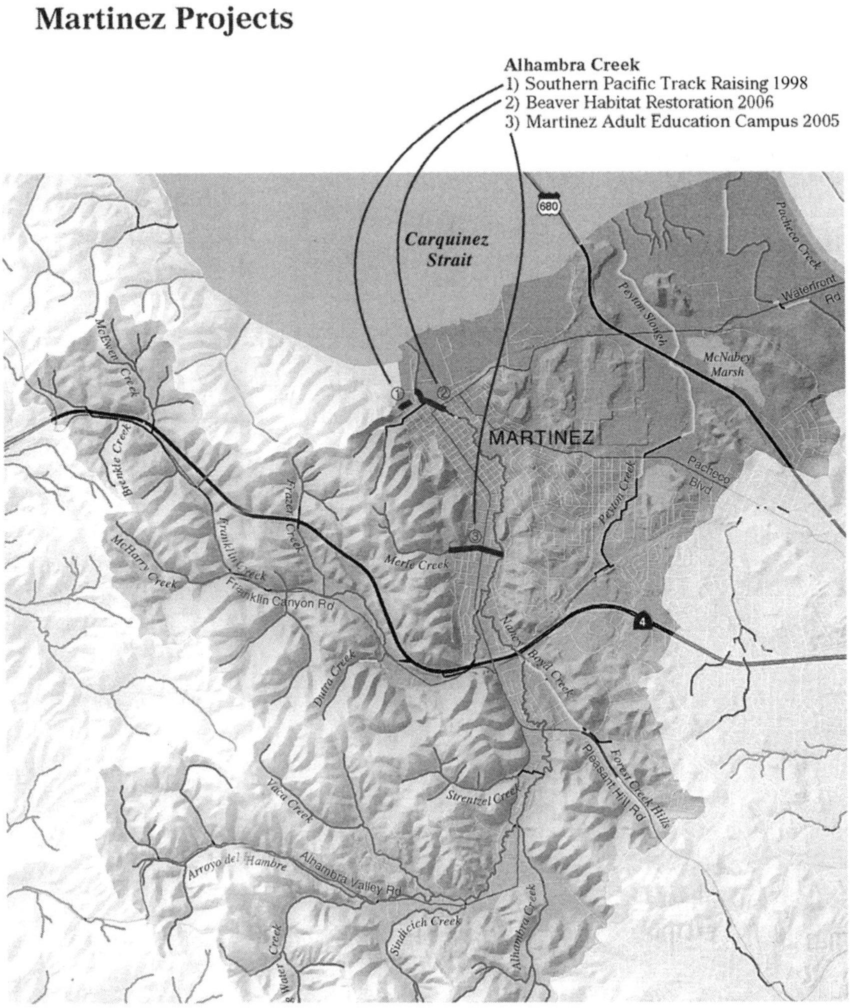

Martinez Projects

Alhambra Creek
1) Southern Pacific Track Raising 1998
2) Beaver Habitat Restoration 2006
3) Martinez Adult Education Campus 2005

FIGURE 1.10 The Alhambra Creek case is located in the city of Martinez, Contra Costa County. *Credit: Lisa Kreishok.*

Creek Watershed Council 2007). It is this lower watershed that creates the setting for the Village Creek case.

Figure 1.9 locates the Baxter Creek restoration cases, which have taken place in El Cerrito and Richmond. The city of El Cerrito forms the boundary for Contra Costa County, situated to the north of Alameda County. Originally founded by refuges from the 1906 San Francisco earthquake, it is a small, residential city with a population of 23,500 contained within 3.7 square miles. It is 53 percent white, 27.3 Asian, and 7.7 percent African American and is generally regarded as a working- and middle-class city, with a per capita income of $32,500. To its north is Richmond, a city of 103,700 within 52.5 square miles. Richmond is 36 percent African American, 31 percent white, and 26 percent Hispanic. Sixteen percent of its residents live below the poverty line, and per capita income is a low $19,800. For many decades, Richmond's local politics were overshadowed by the influence of the Chevron oil refinery, and the city has struggled to contain its crime rates and other issues associated with poverty. In recent years, the innovative police department has contained the crime rates, and Richmond is developing a reputation as a politically and environmentally progressive city independent of local corporate influences.

Martinez, located in figure 1.10, is the seat for the Contra Costa County government. It sits on Carquinez Strait, a busy shipping lane flowing from the delta of the San Joaquin and Sacramento Rivers into San Pablo and San Francisco Bays. Martinez is a small city of 36,000 within 12 square miles. Its population is 70 percent white, 14 percent Hispanic, and 7.7 percent Asian. Its per capita income is $38,000, and the poverty rate in Martinez is a low 5 percent.

References

Belden, Edward. 2008. SWIM Site Indicators, San Gabriel Rivers Watershed. Prepared by Aquatic Bioassay and Consulting for the Los Angeles–San Gabriel Watershed Council.

Bernhardt, Emily S., and Margaret A. Palmer. 2007. "Restoring Streams in an Urbanizing World." *Freshwater Biology* 52:738–751.

Chin, Anne. 2006. "Urban Transformation of River Landscapes in a Global Context." *Geomorphology* 79:460–487.

Clayton, Richard. 2000. "Assessing the Potential for Urban Watershed Restoration." Article 142 in *The Practice of Watershed Protection*, edited by Thomas R. Schueler and Heather K. Holland, 705–711. Ellicott City, MD: Center for Watershed Protection.

Council for Watershed Health. 2014. "Ecology of the Los Angeles River Watershed." *WatershedWise* 15 (4):1–13.

Friends of the Los Angeles River. 1995. *Proposed Flood Control Strategy for the Los Angeles and San Gabriel River System*. Los Angeles: Friends of the Los Angeles River.

Garrett, K. L., ed. 1993. *The Biota of the Los Angeles River*. Los Angeles: Ornithology Department, Natural History Museum of Los Angeles County Foundation.

Hammer, Thomas R. 1972. "Stream Channel Enlargement due to Urbanization." RSRI discussion paper series, no 55. Philadelphia: Regional Science Research Institute.

Leopold, Luna, Reed Huppman, and Andrew Miller. 2005. "Geomorphic Effects of Urbanization in Forty-One Years of Observation." *American Philosophical Society* 149 (3):349–371.

MacAdams, Lewis. 2013. "A River Again in Los Angeles." Los Angeles: Friends of the Los Angeles River.

O'Neill, Thomas, and Gary Hymel. 1994. *All Politics Is Local, and Other Rules of the Game*. New York: Times Books.

Riley, A. L. 2009. "Putting a Price on Riparian Corridors as Water Treatment Facilities." Oakland: California Regional Water Quality Control Board, San Francisco Bay Region.

Schumm, S. A., M. D. Harvey, and C. C. Watson. 1984. *Incised Channels: Morphology, Dynamics and Control*. Littleton, CO: Water Resources Publications.

Simon, A. 1989. "A Model of Channel Response in Disturbed Alluvial Channels." *Earth Surface Processes and Land Forms* 14:11–26.

Stein, E. D., S. Dark, T. Longcore, N. Hall, M. Beland, R. Grossinger, J. Casanova, and M. Sutula. 2007. "Historical Ecology and Landscape Change of the San Gabriel River and Floodplain." Southern California Coastal Water Research Project Technical Report No. 499, Los Angeles.

Surface Water Ambient Monitoring Program. 2008. *Water Quality Monitoring of S.F. Bay Area Streams*. Oakland: California Regional Water Quality Control Board, San Francisco Bay Region.

Watershed Project and Codornices Creek Watershed Council. 2007. "Outreach Plan for the Codornices Creek Watershed." Final Report, February. Richmond, CA: Watershed Project.

Chapter 2

Defining Restoration

The grand old Claremont Hotel in Oakland, California, provided an appropriately engaging setting for an enthusiastic group discussion that would have some historic consequences. The participants were debating how to define and advance a newly evolving concept of environmental restoration. Government employees, scientists, academics, and nonprofit-sector representatives gathered in 1987 in a lounge of this architectural landmark to share their enthusiasm over a new concept for *restoring* the ecological systems that we had been destroying for the last hundred years or more. We were there, in part, to debate a definition of ecological restoration and to help provide a guiding principle for the newly forming Society of Ecological Restoration (SER). Our group reviewed the history of environmental resources management in the United States. In the process, we noted the evolution of terms reflecting a human-centric view of "natural resources" that involved "conserving" resources for efficient use and "preserving" wild areas for the benefit of protecting some of the nation's most spectacular scenery. Our discussion was framed by a conscious effort to avoid the terms *conservation* and *preservation* and their historical connotations in our effort to find the best terms to describe this new movement.

It is hard to imagine that in 1987—not that long ago—the concept of ecological restoration was just beginning to emerge. The 1930s concept of repairing or reversing damages caused by unregulated timber harvest, mining of soils, large-scale air pollution, and other extractive industries entailed a reaction to the Great Plains' Dust Bowl, the collapse of farms, and the filling of reservoirs, rivers, and streams with sediment and debris. The conservation work performed by the Works Progress Administration and Civilian Conservation Corps became our first model for the concept of restoration. This new 1980s movement was different, however, in that its intent was to evolve beyond repairing a damaged landscape through conservation projects such as erosion control, fire control; adoption of new farming techniques; planting hedgerows, buffers, and fire breaks; and "wildlife management" to support hunting and fishing. This new movement also differed from the preservationist movement of the turn of the twentieth century, which supported the national parks movement, and it borrowed from but added another dimension to the development of the conservation sciences that took off starting in the late 1960s and 1970s.

Although we wanted to recognize the recent evolution of the new field of ecological sciences, we knew that we needed to retain people as part of an integrated and complicated ecological system. As our discussion evolved, it became important to us to remove humans as the central reason for resources management. Instead, for the sake of the environments and species involved, we turned our focus on how to return complex and dynamic ecosystems to compensate for the future and current damages that we were committing. William Jordan, a founding member of SER and later editor of its popular journals (*Ecological Restoration*, *Restoration Ecology*), reminded us that ecosystems have evolved in part as a result of human interaction with the environment (he used the influence of human-caused fire as a classic example), and he implored us to include in our definition of restoration the concept that the common person and community needed to be participants in the recovery of ecological systems. He argued that the recovery of natural systems would also represent a recovery of human communities. He pointed out that the benefit to community goes beyond an economic view of the use of and enjoyment of our collective natural resources and includes the re-creation of a "sense of community" and re-establishing a "sense of place." We agreed that the ecological restoration movement should contain the goal to restore functioning environments for a diversity of species to thrive in while also returning a sense of a social and historical locale and identity for the benefit of a generation increasingly disconnected from place by the freeway, strip mall, and car-centric environment.

The first definition of ecological restoration that came out of these brainstorming sessions was eventually published in 1990 by SER. Although this definition has evolved into numerous iterations, it served as a useful early guiding principle for our urban stream restoration practice. It has worn well over time, even as our ongoing deliberations over what restoration should be have become more sophisticated and nuanced:

> Ecological Restoration is the process of intentionally altering a site to establish a defined, indigenous, historic ecosystem. The goal of this process is to emulate the structure, function, diversity and dynamics of the specified ecosystem. (Higgs 1994)

Another early SER developed definition was as follows:

> Ecological restoration is the process of intentionally compensating for damage by humans to the biodiversity and dynamics of indigenous ecosystems by working with and sustaining natural regenerative processes in ways which lead to the re-establishment of sustainable and healthy relationships between nature and culture. (Martinez 1994)

These early definitions of restoration fed ongoing debates about how we really define "indigenous, historic systems." The ecosystems may continue to evolve and cycle into new and old forms, and at what point in history do we select an environment to restore to? There can be conflicts between attaining a condition of greatest biodiversity and encouraging "dynamic and functioning" states. Of course, for those of us attempting to practice restoration in urban environments, we quickly understood that the nature of land use disturbances in our watersheds were never going to be reversed enough to re-create an "indigenous" or historic environment as it existed a few hundred years ago. The enduring aspect of these definitions, however, was the identification of the key variables that should inform any restoration plan.

The first concern was for the structure of forms composing an ecosystem such as stream channels, in-channel forms, floodplains, or different layers of a riparian forest. Examples of ecosystem processes and dynamics can be represented in stream systems through the interactions of riparian forests with the stream channels. Leaves falling from riparian trees become part of the food web; tree roots growing and spreading into the stream bank can create water eddies and encourage sediment transport and depositional forms; tree roots can create and hold the channel boundaries; and trees provide habitat niches, die, fall over, and then produce woody debris that can be major drivers of sediment transport and deposition and of backwatering that establish adjacent wetlands.

The restoration of stream functions can include restoring the ability of the floodplain to frequently store and convey flood flows, store and transport sediment, collect woody debris, and provide rearing refugia for fish. Examples of functions also include groundwater recharge, the absorption of nutrients through the hyporheic zone of the stream bottom, and uptake through plants. By the late 1990s, stream restoration literature became increasingly focused on the need to restore stream processes and functions as well as form, and different perspectives on this issue became part of the discussion that affected the evolution of different schools of restoration. SER published a primer on ecological restoration that helped add greater consistency to the definition of terms used in the discussions, and the terminology of restoration—biotic community, cultural landscapes, biodiversity, ecological processes, ecosystem functions, and ecosystem health, for example—

evolved. SER uses the term *ecological processes* interchangeably with *ecosystem functions*, and literature on rivers often refers to "stream processes." SER defines ecosystem functions or ecological processes as the "dynamic attributes of ecosystems, including interactions among organisms and interactions between organisms and their environments" (Society for Ecological Restoration, International Science and Policy Working Group 2004). The concept of resiliency as an important attribute of restoration was introduced, and it became an objective for restoration projects to be capable of sustaining themselves without regular maintenance or intervention. This sustainability means that the environments would not typically be creating chronic management interventions to address excessive erosion or deposition and would maintain indigenous or native species as opposed to introduced exotic species. The native plants and animal species should thrive, sustain reproducing populations necessary for population stability, and be capable of enduring periodic environmental stress events (such as a fire or flood) that serve to maintain the integrity of the ecosystem (Society for Ecological Restoration, International Science and Policy Working Group 2004).

Restoration Levels

Two different communities of professionals considered how to protect and restore more natural stream systems with greater ecological values and developed two types of nomenclature. The floodplain and flood managers community went through an evolution of thinking that expanded the single-purpose flood control project to river and floodplain modifications as projects that *reduce* flood damages but do not *control* floods and result in multiple environmental as well as risk reduction benefits. The community of ecologists began looking at different levels of ecological functioning that could be achieved in degraded environments.

Many of the cases in this book represent the movement to stop the environmentally destructive, single-purpose flood control projects that were employing channelization, riprap, and concrete in an attempt to make rivers into squares, trapezoids, and rectangles with the idea that these shapes would convey water faster at lower stages. The flood control objectives of these projects were to simplify the stream and floodplain structure and control or eliminate the ecological dynamics and functions that were perceived to be in direct conflict with the ability to "control" floods and maximize developable land. My first book, *Restoring Streams in Cities* (Riley 1998), records the conflict that defines the early origins of many urban river and stream restoration efforts. Efforts were undertaken as early as the 1960s to establish that reducing flood damages and protecting or improving the structure, dynamics, and functions of a stream system, even within the constraints of developed urban environments, were not mutually exclusive objectives.

This clash of flood damage reduction paradigms began to achieve critical mass by the 1980s, defining many stream management efforts continuing to present. By the late 1990s, it was becoming widely accepted that the coequal objectives of flood damage reduction and protecting functional ecosystems could be supported by the same project. In the first decade of the 2000s, more emphasis was applied to the practice of integrating ecological and flood-risk benefits into the same project, recognizing that they are mutually supportive concepts and overcoming the previous paradigm of considering the benefits as mutually exclusive. Project designs have evolved toward having a greater emphasis on the array of "ecosystem services" that these projects should provide. Ecosystem services, for example, are made possible by protecting or restoring stream and floodplain functions. Widely recognized ecosystem services of streams and floodplains are storage of flood flows, improvement of water quality, and support of fish populations. The services include improving water quality through such functions as nutrient uptake, sediment trapping, and temperature control. Protecting or returning stream corridor wildlife such as fish and riparian birds can be achieved with the functions provided by riparian forests, in-stream habitat niches, and variable floodplain flows (Task Force on the Natural and Beneficial Functions of the Floodplain 2002; Kusler 2011).

The evolution of this emphasis on project purposes to broader multiple objectives is illustrated in figure 2.1. The single-purpose flood control projects "to control nature" begin to add on benefits such as trails and recreation to become "multipurpose" projects. The next step was the then-radical idea that protecting rivers and floodplains could provide flood control benefits. The poster child for this concept was the US Army Corps of Engineers Charles River, Massachusetts, Natural Valley Storage flood management project of 1972. This widely celebrated project substituted the construction of a dam with the purchase of a natural floodplain to store the one-in-one-hundred-year flood discharges. Tragically, the federal government essentially abandoned this approach. Floodplain managers and a few flood engineers made concerted efforts in the early 1980s to improve the performance of conventional flood control channels by making them multi-objective. The multi-objective movement recognized that both flood protection and protection of ecological values of rivers and floodplains, recreational opportunities, and community economic development could be complementary. This era also included a concerted effort on the part of some floodplain managers and engineers to apply "nonstructural" approaches to flood risk reduction, which emphasized removing structures and people from the hazard areas through relocations and floodproofing strategies (L. R. Johnston Associates 1989; Association of State Wetland Managers, Association of State Floodplain Managers, and US National Park Service 1991; Kusler and Larson 1993; Association of State Floodplain Managers

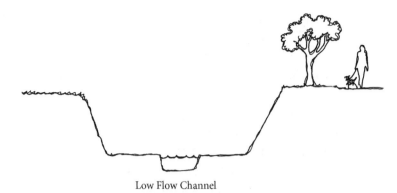

Low Flow Channel

Single and Multi-purpose Flood Control

'Bankfull' Channel

Multi-objective Flood Damage Reduction
Integrating Ecological Function

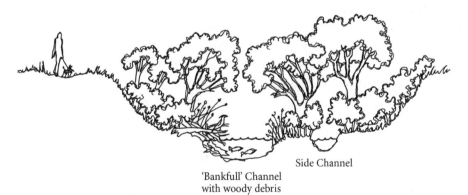

Side Channel

'Bankfull' Channel
with woody debris

Multi-objective Flood Damage Reduction
Ecological Restoration Provides Flood Risk Reduction

FIGURE 2.1 Flood risk reduction projects have evolved to better support river and floodplain functions. *Credit: Lisa Kreishok.*

1996; Federal Interagency Floodplain Management Task Force 1996; National Park Service 1996; Freitag et al. 2009). The multi-objective movement was also a response in part to the large body of influential literature that identified the project performance problems associated with trying to control rivers and floods using the single-purpose engineering methods (Nunnally and Keller 1979; McCarley et al. 1990; Williams 1990; Brookes 1998).

A technical innovation that occurred in flood protection design in the mid-1980s to support the multi-objective project added self-maintaining "bankfull" or "active channels" to the design channel cross section. Sometimes called two-stage channel projects, they feature an equilibrium channel and a floodplain. This advancement recognized the utility of creating a functioning bankfull channel that would better support efficient sediment transport, a deficiency of the old flood control channel designs that were often plagued by excessive sedimentation in overwidened, flat-bottom flood control channels (Williams 1990).

A bankfull channel is a term and concept with origins in the field of fluvial geomorphology. Naturally forming bankfull channels shaped to carry relatively frequent, low-magnitude flood discharges efficiently transport sediment without creating the excessive sedimentation and erosion typical of the engineered trapezoids and rectangles of conventional practice. These bankfull channels, if shaped properly, should not require chronic maintenance, should support in-stream habitat features, and, because they regularly overtop onto floodplain areas, should restore some floodplain functions, including support of fish and wildlife habitat. The positive functions of riparian vegetation were gaining acceptance as a way to moderate stream temperatures, protect water quality, provide needed structural support to stream banks, and reduce the invasive growth of rushes and reeds that could clog channels left completely exposed to the sun.

The next evolution of our thinking about restoring streams in urban settings represented a greater emphasis and appreciation of the ecological functions and services of fully functioning stream riparian areas and floodplains as well as bankfull channels. This awareness entailed a convergence of greater environmental awareness among engineers, the evolution of stronger environmental regulatory programs, the heightened incidence of needing to manage for endangered species, and federal and state funding incentives to restore habitat and greater ecological functionality. The emphasis began to evolve from the concept that flood damage reduction and environmental protection are mutually exclusive concepts to the more positive and proactive concept that flood-risk management can recognize and promote the benefits of floodplain natural resources. Ultimately, preventing impacts to floodplain environments supports efforts to protect the public from flood risk (Plasencia and Monday 2009). Floodplain management theory came full circle back to the concepts that enabled the Charles River Natural Valley Storage Project.

A new generation of literature that assigned values to ecosystems services provided by rivers and floodplains, sometimes in monetary terms, was followed by a new generation of Army Corps projects in which ecosystem services were integrated into some of their project cost-effectiveness evaluations. A 5,000-foot reach of Wildcat Creek, an urban stream in Richmond, California, for example, provides water quality benefits equivalent to those of the Santa Monica, California, Urban Runoff Recycling Facility shown in figure 2.2. The Santa Monica Plant was constructed for about $15 million and requires about $200,000 annual maintenance to treat stormwater before it enters Santa Monica Bay. In comparison, the Wildcat Creek restoration of ecological functions was accomplished through one of the more expensive projects involving federal participation, yet it still cost less for similar benefits as the Santa Monica facility. Annualized over a fifty-year period, the Santa Monica facility costs are $1.3 million per year compared to $967,000 a year for the natural systems restoration for water quality benefits that were achieved from the Wildcat Creek flood reduction project. A more typical stream restoration cost for a project of this size, amortized over a fifty-year period, would be $227,000 (2008 dollar values). These comparisons are only for water quality benefits. The natural stream system that is part of the Wildcat Creek project also provides fish and wildlife habitat services that the plant cannot produce (Riley 2009). Ecosystem services of functioning streams and rivers are now widely being monetized in a new era of economic studies (DeGraff and Batker 2011).

The ladder in figure 2.3 represents the different levels of restoration projects in the context of how river ecologists view restoration. The highest rung denotes the most ambitious level for a restoration practitioner to achieve: the ideal of returning a damaged landscape into an "indigenous, historic" ecosystem. This objective is ideal but largely unattainable for most circumstances, urban or rural, because the watersheds have changed too much to support a past environment. Ecological restoration makes up the next rung on the ladder, where the objective is to return as much ecological structure, dynamics, and functions as is achievable. This level results in a functioning ecosystem with measureable improvements in aquatic and terrestrial wildlife made possible by a dynamic, unrestricted stream channel, floodplain, and riparian corridor.

The functional rung on the ladder in figure 2.3 recognizes that the highest levels of biological recovery may face insurmountable or long-term challenges because of land use changes or constraints. Functional restoration recognizes, however, that many of the basic processes of streams can be returned, such as the "messy" businesses of erosion, deposition, floodplain flooding, tree and root growth

FIGURE 2.2 Functioning streams provide quantifiable ecosystem services equivalent to constructed treatment plants such as the Santa Monica Urban Runoff Recycling Facility.

Restoration Levels

- Historical Restoration
- Ecological Restoration
- Functional Restoration
- Enhancement of Controlled Channels

FIGURE 2.3 Restoration projects can support different levels of ecological functioning.

and tree fall, movement and sorting of bed loads, and channel slope and planform changes that combine to support river habitat. The historic occupation of the site by native fish species may, for example, be constrained by problems elsewhere in the watershed caused by migration barriers, unresolved water pollution, or runoff hydrograph changes, but the stream is no longer prevented from performing basic physical processes such as sediment transport and deposition, erosion, vegetative growth, channel planform changes, bed load sorting, and riffle, pool, and step development that are associated with functioning habitat. An example of a functional environment is a stream that has been leveed from its historic floodplain but still has sufficient floodplain to support a meandering channel, riparian corridor, and in-stream habitat.

In my experience, what was often represented as "restoration" was the condition represented by the lowest rung of the ladder, "enhancement of controlled channels." These so called "restoration" projects typically took the form of an engineered (often straightened) channel, employing grade controls, rocked banks, nursery container plants added as an aesthetic enhancement, and a trail or pocket park added as a social amenity. See the multipurpose paradigm in fig. 2.1. In these cases, the planting pallet may add some plant species diversity, and people may enjoy the flowering plants, but the stream channel sediment transport and deposition, channel meandering, vegetative growth, and channel dynamics are gone or greatly compromised. We have just decorated a corpse. The pretty plant placed on the bank is just that. If it does not exist within the context of the complex forms and processes at work in an ecological system, it has no other function than to beautify the landscape as a garden would.

Some restoration projects are focused on removing invasive, exotic plants and replacing them with the native riparian plants that may have been typical and regular occupants before land use changes or invasions by weeds. These projects may increase the functioning of a stream corridor in terms of erosion control, shade, and habitat for native fauna if the replanting re-creates an appropriate reference condition for the location of the site and meets criteria for being self-sustaining and minimizing the need for human intervention. Such a project may entail the re-creation of multiple layers and canopies, dense brush and undergrowth, or other features that characterize a functioning riparian plant community. These kinds of projects can have excellent value, but they are vulnerable to falling into the landscape gardening category if the emphasis is on aesthetics rather than ecosystem function and recreation.

I discovered a similar scheme for describing the universe of "restoration" practices in the book *River Channel Restoration* by Andrew Brookes and F. Douglas Shields Jr. (1996). They similarly describe the largely unattainable "full restoration" as denoting the complete and functional return to a predisturbance state.

They use the term *rehabilitation* to represent partial return to a predisturbance ecosystem structure and functions. They describe *enhancement* as any improvement in a structural or functional attribute, but not representing an environment of a predisturbance condition. They also list the term *creation*, which is used to represent the construction of a new alternative ecosystem that did not previously exist at the site. A third term, *naturalization*, represents a hybrid of some of these terms. Naturalization recognizes that human use and interaction are components of the current "natural environment," and naturalization has the objective of "determining morphological and ecological configurations that are compatible with contemporary magnitudes and rates of fluvial processes" (Brookes and Shields 1996, 4). Rehabilitation, creation, and naturalization as described here fit well with the second and third rungs on our restoration ladder in figure 2.3 that correspond to functional restoration and ecological restoration. Enhancement fits well with the ladder's bottom rung, and full restoration correlates well with the top rung. The restoration levels in our case studies distinguish more between achieving recovery of geomorphic and riparian processes (functional) and achieving biological and ecological benefits as a result.

It would almost be possible to just write a book about the definitions of terms associated with the management and restoration of streams and floodplains. A federal manual, *Stream Corridor Restoration*, developed in 1989 and representing a collaborative effort of fifteen federal agencies, became a widely used and accepted resource on restoration concepts, practices, and terminology (Federal Interagency Stream Restoration Working Group 1998). The US Environmental Protection Agency (EPA) has tried to standardize definitions for restoration, rehabilitation, enhancement, and creation because these terms are used within a legal enforcement system to protect wetlands and water quality (Compensatory Mitigation for Losses of Aquatic Resources 2008). The EPA refers to preservation as removing a threat to or preventing the decline of a resource. It uses the terms *creation* or *establishment* to refer to developing an aquatic resource that did not previously exist at a site. *Re-establishment* means returning a former aquatic resource that existed at the site both in form and function. *Rehabilitation* is used to describe a gain of function by the "repair" of a natural or historic function at a degraded aquatic site, but not a gain in wetland area. *Enhancement* is defined as achieving a gain in a selected resource function, but it may also lead to a decline in other resources' functions.

The term *natural beneficial functions of floodplains* is favored by floodplain management professionals. It describes the natural resources of floodplains, which are identified in federal publications as natural flood and erosion control, water quality maintenance, groundwater recharge, biological productivity, fish and wildlife habitats, harvest of wild and cultivated products, recreational opportunities, and scientific education and study (Task Force on the Natural and Beneficial Functions of the Floodplain 2002). The term *beneficial uses* is commonly used in the water quality regulatory and floodplain management circles and usually contains lists of uses described as consumptive (e.g., hunting, fishing, forestry, mining, groundwater withdrawal) and nonconsumptive (e.g., recreation, transport, aesthetics).

All the restoration cases in this book were intended to achieve the functional and ecological restoration levels in figure 2.3. The purpose for my involvement in the projects described here was to demonstrate that these kinds of projects are possible in urban settings. I also wanted to motivate those operating at the lowest rung of the ladder to increase their benefit-cost ratios and delivery of ecosystem services and reach toward the higher rungs of the ladder.

Figures 2.4 and 2.5 show functional levels of restoration. The Wildcat Creek channel in figure 2.4 is an ecologically functioning channel that was part of a multi-objective flood control project constructed between 1986 and 2000 by the Army Corps of Engineers and Contra Costa County in Richmond, California. The creek was degraded from vegetation removal and some sporadic dredging by property owners anxious to reduce frequent flood damages. The multi-objective project provides for an ecologically functioning channel and riparian forest protected as an unmaintained feature of the project. The channel is a dense riparian corridor and habitat for the native anadromous steelhead trout. Figure 2.5 shows the Wildcat Creek floodplain that represents more of a functional restoration feature as opposed to ecological restoration. The floodplain provides for "natural" storage and conveyance of flood flows. Its ecological function is limited by occasional maintenance that involves mowing of the floodplain vegetation when elevation surveys and flood-level data indicate that design floodplain capacity is impacted beyond the project's flood-protection objective.

Figure 2.6 shows a functional restoration project attained through the daylighting of Baxter Creek in a median strip between two streets in El Cerrito, California. The constructed step pool channel has unconstrained and mobile cobbles and boulders. The riparian corridor, established with soil bioengineering, is a wild oasis in a densely populated neighborhood in which deer and mountain lion have been observed. The project is constrained between the two streets and two culverts and thus falls in the functional restoration category in figure 2.3.

Another example of a functional project is the conversion of the Santa Rosa Creek from a concrete channel. Santa Rosa Creek, in the downtown historic district of Santa Rosa, was channelized as a single-purpose flood control project by the Soil Conservation Service and Army Corps of Engineers in the early 1960s (fig. 2.7). Meanders were filled, and the channel was constructed as a trapezoidal flood conveyance facility, with rock boulders cemented into the bottom in a vain

FIGURE 2.4 The Wildcat Creek, Richmond, multi-objective flood risk reduction project is an ecological restoration project completed between 1986 and 2000.

FIGURE 2.5 The Wildcat Creek project improves ecological functions associated with the channel, floodplain, and riparian forest.

FIGURE 2.6 The Baxter Creek daylighting project, which removed a culvert in a median strip between two streets in El Cerrito in 1996, is an example of functional restoration.

attempt to provide protection from high stream flows for fish. The restoration planning began in 1993 with a new master plan for the creek to help revitalize the downtown environment and reverse the intolerable conditions for the native anadromous fish, which were subjected to high water temperatures and difficult passage conditions. The city has since accomplished a series of phased restoration projects that began in 1999 to remove the concrete from the channel bottom and sides (Owens Viani 2003).

FIGURE 2.7 The Santa Rosa Creek project located in downtown Santa Rosa, Sonoma County, is an example of a single-purpose flood control project constructed in the 1960s.

The Santa Rosa Creek project is an example of a functional restoration in which the stream now erodes, transports, and deposits sediment; supports a narrow riparian corridor; and provides functioning habitat and passage for the native anadromous fish within the existing rights-of-way of the original flood control project (fig. 2.8). It also is multi-objective, protecting the original flood risk reduction function of the original design discharge of 10,500 cubic feet per second as shown in figure 2.9.

Figures 2.10 and 2.11 show projects that were funded as "stream restoration" projects by the State of California, but they represent the category of enhanced controlled channels. Cerrito Creek, shown in figure 2.10, was once a steelhead trout stream but was straightened and locked between large boulders in 2003, preventing the processes of meandering, erosion, and deposition characteristic of a functioning riparian corridor to provide shade, water quality benefits, and habitat. The Brush Creek project shown in figure 2.11 was intended to restore a trapezoidal flood control channel to improve fish habitat in 1999. The project's plans involved an artist-drawn low-flow channel that was not designed to represent ecologically functioning bankfull dimensions. The channel is locked in place by large boulders, and the planted trees do not provide near-channel or floodplain ecological functions. The native anadromous fish disappeared from the enhanced

FIGURE 2.8 Removing the concrete bottom of the Santa Rosa flood control channel in 1999–2000 allowed the return of a limited functioning riparian corridor, channel sediment transport and deposition, a channel-floodplain connection, and anadromous fish passage.

FIGURE 2.9 Restoration of ecological function to Santa Rosa Creek did not compromise the flood risk reduction objective. This photo shows the channel containing the large San Francisco north bay flood of 2005–2006.

controlled channel because the project increased stream temperatures. Figure 2.11 misleads many into thinking that this project represents a functional restoration because it contains some of the features often seen in restoration work, such as a low-flow channel, root wads, erosion control fabric, and staked, planted trees.

Form and Function

Architects are taught that form should follow the functions desired for landscapes or structures under design. There is quite a bit of discussion in today's restoration literature about the question of whether restoration of landscape forms, such as bankfull channels and floodplains, can qualify as restoration. Questions have evolved over the linkages between form and function. Can we copy more "natural" stream forms to use as a basis of stream restoration design? A strength of process-based restoration is that it focuses on addressing the sources of environmental degradation. Should we only use information about stream processes to design projects? Do we use both form and process information? Can it be argued that using form-based design is necessarily in conflict with the restoration of processes? Yet another version of the debate about form versus function asks, Should we design stream forms at the expense of the recovery of stream processes so as to support a popular view of landscape aesthetics, even while compromising ecological value?

Conversations sometimes reflect strong biases for applying process-only information about rivers, such as sediment transport characteristics, as opposed to using field data collected on river forms to inform restoration. Some concern may be particularly directed at restoration projects that try to emulate a stream type representing a desired equilibrium that has occurred in the recent past. Some argue that this stream type may no longer sustain itself because watershed conditions have changed too much to sustain a landscape of the past. Others argue that relationships among channel forms and processes are critical and practical design tools.

The related issue about restoration approaches is that a restoration project based on the general public's perception of river health—that of a tidy greensward park—can preclude the public's understanding of the importance of river environment processes such as flooding and erosion that maintain the "messy" river characteristics responsible for ecological functioning. A segment of a river may meet many people's expectations of a healthy river if the water is clear and the stream banks are not eroding. If this "tidy" appearance comes at the expense of ecological functioning, however, the hydrologic and geomorphic processes may no longer create and maintain the disturbance regime necessary to support ecosystem integrity (Wohl 2005).

FIGURE 2.10 The Cerrito Creek Project in El Cerrito represents an enhanced, controlled channel with streamside gardening and was completed between 2000 and 2002.

FIGURE 2.11 The 1999 Brush Creek project in the Santa Rosa area represents an enhanced-controlled channel, locking in the channel with large boulders and adding landscaping trees that do not perform ecological functions such as temperature control for the stream system.

Recent research on some of the pioneering geomorphic studies of streams by M. Gordon Wolman and Luna Leopold points out that although their fundamental insights on river dynamics and related form certainly hold up over time, some of the streams and rivers from which they collected their data do not necessarily represent "natural" or nonimpacted environments (Montgomery 2008; Walter and Merritts 2008). Although it is important to understand the context for some of the science collected from human-impacted landscapes, doing so does not negate the value of information we gain from measuring landscape evolutions over time under different land use conditions and the relationships of river forms to watershed conditions and processes. Useful information to guide restoration objectives can come from a range of sources, including knowledge of historic landscapes, the use of reference sites, and their conditions, reflecting either the recent past or current healthy landscapes. Restricting ourselves to either landscape-form information or process-based information, which are obviously related, does not make good sense because we need a full array of perspectives and methods to understand rivers and guide their recovery.

By 2006, the discussion about restoration objectives achieved a new level through the efforts of scientists as recorded by the Army Corps of Engineers Environmental Research Development Center (Fischenich 2006). Fifteen critical functions provided by stream, riparian, and floodplain corridors were defined by a committee of scientists, engineers, and practitioners to serve as a basis for stream assessment, design, and management. The message from this report is that quality stream ecosystems are a product of healthy watersheds, wide and relatively continuous riparian areas, active floodplains, suitable channel dimensions for the prevailing conditions, and an appropriate level of diversity and dynamics. Because efforts to restore degraded streams can be ineffective if they fail to address the underlying processes that create and maintain stream functions, an important part of our "job" is to recognize and define these functions. A summary of these fifteen critical functions is listed in table 2.1.

The optimum way to assess a restoration project is to use measurements and observations that can track changes in both form and function over time. The level of detail and completion of project evaluations are, unfortunately, constrained by the cost of information collection and analysis as well as the time it takes to discover how an environment may be evolving. The case studies provided in this book are a start in recording the changes in structure, dynamics, and functions of created environments over time.

Restoration Objectives for Case Studies

My own definition of restoration and what constitutes a successful restoration project evolved over a twenty-five-year period of practice, similar to the evolution that

TABLE 2.1.

Fifteen critical functions of streams, riparian, and floodplain environment processes, descriptions, and indicators

1	Hydrodynamic character	Flow conditions and fluctuations at different seasons to support the biotic environment; flood flows on active floodplains
2	Stream evolution processes	Promotion of changes necessary to maintain diversity and succession; complexity of channel forms and flow; abundance and distribution of pioneer species as a succession to a diversity of quantity, densities, and ages of vegetation types
3	Surface water storage processes	Storage of high flood flows; replenishes soil moisture; pathways for fish, low-velocity habitats; presence of floodplain and wetland features; riparian debris and detrital accumulations
4	Sediment continuity	Erosion, transport, and depositional processes, substrate sorting; establishment and succession of riparian habitats; nutrient cycling; floodplain deposits; channel planform and bed sediment character and fluctuations
5	Riparian succession	Changes in vegetation structure, age, diversity, maturity; presence of pioneer species, varied age classes, diversity; new sediment deposition, large woody debris recruitment
6	Energy management	Spatial and temporal variability in cross section, grade, and resistance; habitat creation; changes in physical channel features over time
7	Substrate and structural processes	Channels and riparian zones provide stream architectural structure that supports resilient diverse habitats
8	Quality and quantity of sediments	Organisms are often dependent on specific sediment regimes, sediment yield and character, channel, bank, pool and bar forms; distribution, abundance, and diversity of biota
9	Biological communities and processes	Diverse assemblages of native species, natural reproduction and long-term biotic persistence; changes in condition of individuals or populations
10	Surface-subsurface water exchange	Bidirectional flow from open channel to subsurface soils; exchange of chemicals, nutrients, and water; subsurface water storage, base and seasonal flows; invertebrates in hyporheic zone; floodplain for groundwater recharge
11	Water and soil quality	Trap, retain, and remove particulate and dissolved constituents; regulate chemical and nutrient cycles; control pathogens; plant, fish, invertebrate density, diversity, and distribution; water quality parameters (e.g. dissolved oxygen, temperature, pH, nitrogen, phosphorus)
12	Landscape pathways	Corridors for plant and animal migration; source areas for maintaining population equilibrium of plant and animal species
13	Trophic structures and processes	Promotes growth and reproduction of biotic communities across trophic scales; presence of a variety of nutrients and organisms to convert carbon, nitrogen, and phosphorus between forms; aquatic and riparian vegetation density, biomass production, large woody debris frequency and density
14	Chemical processes and nutrient cycles	Acquisition, breakdown, storage, conversion, and transformation of nutrients; riparian vegetation composition and vigor; seasonal debris in riparian area
15	Necessary habitats for all life cycles	Basic food, air, light, water, shelter needs; reproduction, migration, temporal habitats during periods of population stress; presence and complexity of habitat features

Source: Adapted from Fischenich 2006.

occurred with other practitioners and academics. The six case studies use success criteria described in recent literature to evaluate projects conceived decades ago, years before the thinking about evaluation criteria was as well evolved as it is today. A paper written with the input of twenty-two authors (Palmer et al. 2005) lays out a context for evaluating the success or lack of success of a restoration project. In this book, the five categories of success from this paper, along with two I have added to better reflect the value of projects to instruct and provide community benefit, will be used. The sixth category is added to better capture that we are also "practicing" restoration to learn from our experiences in restoration planning, design, and construction. The seventh category is added to emphasize our objective to deliver tangible community benefits such as job training and creation, neighborhood safety, and improvement. It also captures the spirit of the meeting we had at the Claremont Hotel back in 1987 to recognize the link between humans and the environment. The seven categories are each discussed in turn.

1. Create an Ecologically Dynamic Environment
The project needs to describe and accomplish an ecologically dynamic state and recognize that there are ranges in the variables making up a stream. This process can include historic research, not necessarily to replicate a historic landscape, but to point out ecological potentials and the irreversible changes and constraints on future potentials. Reference sites can include environments that represent a state of recovery as well as nearby restoration projects that have adjusted to the environmental conditions acting on them. Analytical or process-based models and empirical relationships among hydrologic, hydraulic, and biological variables are combined to guide restoration. The plan should consider local and watershed processes and stressors and should move the stream to the least degraded and ecologically dynamic state possible.

2. Improve Ecological Conditions
It should be possible to measure some improvements in ecological condition using indicators such as water quality, increases in the populations of target species, percentage of native versus nonnative species, extent of increased riparian vegetation, bioassessment index improvement, and improvements in addressing limiting factors for a given species of life stage. Reach-level improvements can be evaluated within the context of whether they are part of multiple projects in the same watershed.

3. Increase Resiliency
The environmental system created should require minimal ongoing intervention to sustain itself and have the capacity to recover from natural disturbances such as floods and fires.

4. Do No Harm
The project interventions should not cause irreversible damage or lasting harm to the ecological properties of the ecosystem.

5. Do Ecological Assessments
An ecological assessment is conducted on the basis of a before-and-after conditions basis or on a treatment-control basis. Well-documented projects reflecting mistakes, lost opportunities, or results falling short of restoration objectives may ultimately contribute more to restoration science than projects that appear to fulfill all predictions of restoration objectives.

6. Create Learning about Restoration Planning, Design, and Construction for the Future
The project should create a learning opportunity to understand more about restoration design and planning. Included here are evaluations of whether planning methods successfully involved the relevant stakeholders and whether the science involved in informing restoration was well understood by designers, decision makers, and regulators and communicated to a broad stakeholder group. Does a project overcome obstacles to realizing functional restoration? What are the results of different design strategies, construction, and installation methods?

7. Create Community Benefits
The project should be perceived as an amenity by the community in which it is located and should address social and economic needs such as education, job creation and training, and neighborhood quality-of-life improvements. Projects should be evaluated on the basis of whether community and social benefits are pitted against realizing functional restoration or whether social and environmental objectives are communicated in the way that these kinds of conflicts are avoided or minimized.

Schools of Restoration

Restoration has also been defined by the different traditions and perspectives associated with fluvial geomorphology, hydraulic engineering, ecology, plant ecology, wildlife biology, and floodplain management. These disciplines have produced distinguishable schools of restoration based on how they have been applied within a newly forming practice involving multiple sciences to restore the environment. Disagreements over what methods produce better results, such as the form versus function debates mentioned above, have almost come to blows. For instance, many engineering professions have favored analytical hydraulic modeling over the use of field data based empirical relationships that govern channel shapes and dimensions. Some people believe that riverscapes should recover on their own and that we should minimize human interference with restoration actions. Still others believe that restoration will come about only, or chiefly, by stormwater control projects that change the hydrology of a watershed. Many fish, wildlife, and plant scientists have different perspectives on whether to emphasize population recoveries or instead focus on protecting and increasing genetic or species diversity.

The *empirical school* evolved from the field of fluvial geomorphology that pioneered the assessment of field observations from rivers to develop relationships among river and floodplain forms and processes. The field data can help develop regional or watershed-scale relationships among some of the variables making up river systems, such as watershed drainage areas, rainfall, channel shapes, and discharges. The assessments can be used as reference information on "balanced" river dimensions for application by practitioners of river management and restoration; to provide initial evaluations of how impacted a stream system may be compared to other, less impacted areas or more fully functioning streams; or to indicate how riverine landscapes are changing over time due to climate change or other causes. A classic paper on hydraulic geometry of streams (Leopold and Maddock 1953) was a pioneer in the empirical school. The restoration tools applied from this school in our case studies include the use of regional restoration curves to estimate channel dimensions and the application of hydraulic geometry relations between channel dimensions and meanders. The detractors of this school often believe that these tools are not suited to urban environments undergoing various stages of adjustments and degradation and that urban incised channels make it too difficult to determine the dimensions of bankfull or active channels, thereby making correlations difficult. They warn that information from regional curves needs to be applied to similarly situated environments and not applied to too large a geographic region. The most extreme version opposing the application of this school is to doubt its scientific basis that equilibrium channels can represent a stable stream form over a significant period of time (Knighton 1998; Natural Resources Conservation Service 2007; Shields et al. 2008).

The *analytical school* is the realm of the hydraulic engineer who uses quantitative models of river processes such as continuity, flow resistance, and sediment transport to characterize the relationships between discharges, channel slopes, shapes, and sediment transport (Shields et al. 2008). The models are used to estimate whether a river project design will have effective but not excessive sediment transport, to estimate forces acting on streambeds and stream banks, to determine whether floodplains and channels have sufficient area to transport flood flows, and to indicate flood-flow elevations for different discharges. Modeling is now applied to fish-passage design projects to identify velocities and depth of flow under different conditions. Information from both the empirical and analytical schools can be used to cross-check results from the others in estimating factors such as flood stages and channel-forming discharges. The well-accepted mathematical equations and quantitative outputs create a strong following for this school. The detractors of the analytical school point out that the models simplify river environments because they cannot represent all the variables acting on rivers. Geomorphologists advise that approximately fifteen variables are required to fully describe a dynamic stream system, with nine of these variables remaining unknown. This oversimplification has resulted in a generation of poorly performing flood control channels that did not take into account sediment transport and deposition. Furthermore, there is the difficulty in collecting enough field data to calibrate models so that they adequately represent the natural conditions (Hey 1988; Thorne, Hey, and Newson 1997; Soar and Thorne 2001).

Stream evolution represents another process-oriented school. Here, sketches are used to help identify stream responses over time to changes in watershed sediment supplies, discharges, channel slopes, and vegetation. The classic references for this stream evolution school relate to the work of various river scientists (Lane 1955; Schumm, Harvey and Watson 1984; Simon 1989). Channel evolution models are particularly useful for predicting how a stream and its floodplain may change over time in response to changes made to the channel or watershed. These models capture watershed processes in simple drawings, which can be very useful when assessing the causes of stream system instabilities and describing the processes that act on a riverscape. The evolution models are improving over time, with more variables and scenarios described to represent stream responses to different channel and watershed changes. The stream evolution models may oversimplify the numerous variables acting on stream systems. The currently available models also do not cover the range of the wide variety of different stream types, and they leave out the element of time in which the processes are expected to evolve (Riley 2003; Hey 2007; Cluer and Thorne 2014).

The use of *river classification* systems is another developing school of river and floodplain assessment. This school applies the concept that stream reaches can be grouped into major types of riverine landscapes to develop reference data for managing or restoring similar stream types. The classifications use descriptions of landscape forms, and these forms may provide data that inform river processes. The use of classifications produces the most differences of opinion among professionals but will likely continue to evolve as a part of the river assessment management field. In the most widely applied stream classification system (Rosgen 1994), reference reach information is applied so as to inform restoration objectives and designs for degraded streams. Another widely known classification system (Montgomery and Buffington 1993, 1998) involves high-gradient cascading and step pool streams and other riverine environments typical to the Pacific Northwest.

Yet another classification system uses a "river styles" framework that focuses on valley settings and geomorphic units within a valley segment (Brierley and Fryirs 2005). This classification emphasizes the role of watershed processes upstream on downstream reach conditions. An example of a simple classification system is the description of three basic channel patterns: straight, meandering, and braided (Leopold and Wolman 1957). This river pattern–based classification spun off a variety of classifications for alluvial channels (Roni and Beechie 2013).

The widespread adoption of stormwater regulatory programs has been an influential aspect of riverine assessment and a "driver" for watershed management activities and projects. It stands out as a school of its own because it can advocate a runoff-control-only approach to restoring stream functions. For example, the Center for Watershed Protection has led in developing percent-watershed impermeable indicators as a way of capturing the impacts of urbanization on the health of stream systems (Schueler 1994; Schueler and Holland 2000). Other researchers have derived different land use and water management indicators that they consider more useful in characterizing the ability or limitations of urban streams to function as habitats in developed areas, raising issues about the utility of imperviousness as a key indicator of restoration potential (Roesner and Bledsoe 2003). Urban stormwater management is still in its first decades, and there are mixed conclusions about the actual causes of degradation associated with the increases in urban runoff, such as the differing levels of impacts associated with increased shear stresses and volumes of stream flows, channel incision, pesticide levels, and loss of vegetative cover. There is also a range of results from different urban stormwater management strategies, such as the new use of "green streets" and "landscape-based" stormwater infiltration projects.

The area of stormwater management is grouped with the *passive school* of river restoration. Here the emphasis is placed on improving watershed conditions through changing the rates and magnitudes of runoff and increasing or decreasing sediment supplies to develop greater balance among sediment discharges and flows. Proponents of the passive school are often opposed to or discourage actions that change channel or floodplain dimensions, preferring that nature make the changes on its own after watershed conditions have been improved. Process-based restoration is described as taking the actions needed to address the primary causes of ecosystem degradation and recognizes that reach-scale processes as well as regional landscape and watershed-scale processes can be part of the recovery strategies. Examples of reach-scale measures that can return processes are the reintroduction of woody debris and channel reinforcement with vegetation (Beechie et al. 2010).

Another major area of restoration practice that produces different schools of restoration is the application of *fish, wildlife, and plant community ecology* to inform re-establishing habitat functions in stream corridors and improving watershed conditions and processes to support both native riparian communities and the aquatic habitat. The beginnings of the movement toward restoring rivers in the United States can be traced to American Fisheries Society efforts in the 1930s to conserve and improve in-stream habitat that emphasized creation of pool habitats, erosion control, and migration aids. Evaluations of these in-stream structures found them to have important limitations, and these practices evolved to more sophisticated methods for habitat recovery in which biologists began to assess which limiting factors were keeping potential population recovery low. This analysis was then used to better inform in-stream habitat modifications. Watershed-scale assessments involving hydrology, sediment, habitat connectivity, riparian conditions, and recovery of in-stream and floodplain complexity added to the sophistication of restoration planning and approaches. Watershed processes as well as conditions are now being evaluated. Applying a regional perspective to fish population management grew from the realization that to maximize a network of habitat refugia within and among watersheds to support genetic variability, large-scale recovery planning needs to be added to restoration strategies.

The two major schools evolving from the fields of fish, wildlife, and plant community biology are the *population abundance school*, which is most focused on increasing populations of a particular species, and the *genetic diversity school*, which is most focused on protecting or increasing the genetic diversity of populations. Although practitioners will integrate both concerns into their practice, the different schools often represent a significant difference of emphasis. Hatcheries, limiting-factor analysis projects, and reach-level habitat projects most often represent the objective to increase fish populations or abundance for particular species. The regional and watershed-level strategies often focus on increasing options for all life stages of fish and accommodate multiple aquatic species in a diversity of protected or restored environments; they also most often represent an emphasis on genetic diversity (Beechie et al. 2008).

Restoration of plant communities can likewise be characterized by active replanting projects or the return of more natural hydrologic processes to allow natural regeneration of riparian species. Vegetation management approaches may emphasize watershed-scale activities such as fire management activities, erosion control, and protection of riparian buffers; protecting connected corridors for wildlife refugia; or the more active replanting of new riparian reaches. The restoration practices of revegetation represent four different practicing schools. One is the *landscape design school*, which emphasizes the traditional principles of landscape architecture practice. Here, the focus of a project is meeting social and other developed-site programmatic needs along the stream corridor, including agricultural and urban land uses. The landscape design school generally uses what are considered conventional planting design and installation methods, including heavy reliance on nursery-grown container stocks of plants. In contrast, the *soil bioengineering and plant community functions school* emphasizes plant community restoration through the planting of a few pioneer, early succession species in bundled assemblages of plant material cut from nearby growing native stock. The plant material collected as cuttings from existing riparian corridors stabilizes a degraded system and forms the basis for ecological recovery. This evolving practice combines plant physiology and ecology with principles of mechanical engineering. This school emphasizes achieving balanced stream systems that avoid excessive erosion and the return of a quickly functioning habitat with long-term resilience. This strategy can employ adding a greater diversity of riparian species over time, if appropriate (Gray and Sotir 1996).

A third school employs *horticulturally based restoration* in which native plant stock is grown and planted out in floodplains and riverbanks with a conscious effort to match the species with flood elevations and soils. In this type of restoration, multilayered forest levels—sometimes grouped in mosaics to function as cover, shelter, and food for both riparian bird and fish species—are installed. A fourth and related school is *process restoration*, with a focus on returning greater hydrologic variability characteristic of more natural flood regimes of streams that have been affected by the control of flows by dams, reservoirs, levees, and berms. This practice can return the flows to floodplains, help maintain floodplain environments and processes over longer inundation periods, and in some cases allow large magnitude floods to return to "reset" new floodplain environments. It enables the "volunteer" re-establishment of plant communities that reinhabit an environment on their own if the conditions are appropriate.

Each of these schools has produced different bodies of literature and added a rich diversity of perspectives to the evolving field of restoration. They represent different professional emphases and training within the broad, integrated field of restoration. In some instances, the result is differences about the most "valid" approach to restoration project design. The training, experience, and comfort levels of professionals within the different schools can sometimes have the effect of limiting river managers to using one or two paradigms at the exclusion of others. The different perspectives and acceptance of the schools may, at times, result in conflicts among practitioners or between practitioners, academics, and regulators. Private clients contracting for stream management services, government agencies, and the public can unwittingly get caught in the crosswinds of the different schools, and sorting out the causes of the conflicts can be difficult. The cases in this book represent design processes that integrate the use of all or many of these different schools and therefore illustrate how the various disciplines and traditions of river study can complement rather than conflict with one another.

References

Association of State Floodplain Managers. 1996. "Using Multi-Objective Management to Reduce Flood Losses in Your Watershed." Guidebook prepared for the Environmental Protection Agency, Washington, DC.

Association of State Wetland Managers, Association of State Floodplain Managers, and US National Park Service. 1991. "A Casebook in Managing Rives for Multiple Uses." Washington, DC: National Park Service.

Beechie, T., G. Press, and P. Roni. 2008. "Setting River Restoration Priorities: A Review of Approaches and General Protocol for Identifying and Prioritizing Actions." *North American Journal of Fisheries Management* 28:891–905.

Beechie, Timothy, David Sear, Julian Olden, George Press, John Buffington, Hamish Moir, Philip Roni, and Michael Pollock. 2010. "Process-Based Principles for Restoring River Ecosystems." *Bioscience* 60 (3):209–222.

Brierley, G. J., and K. A. Fryirs. 2005. *Geomorphology and River Management: Application of the River Styles Framework.* Oxford: Blackwell.

Brookes, Andrew. 1998. *Channelized Rivers: Perspectives for Environmental Management.* New York: Wiley.

Brookes, Andrew, and F. Douglas Shields Jr. 1996. *River Channel Restoration: Guiding Principles for Sustainable Projects.* Chichester, UK: Wiley.

Brookes, Andrew, and F. Douglas Shields Jr. 1996. "Towards an Approach to Sustainable River Restoration." In *River Channel Restoration: Guiding Principles for Sustainable Projects*, edited by Andrew Brooks and F. Douglas Shields Jr., 385–402. Chichester, UK: Wiley.

Cluer, B., and C. Thorne. 2014. "A Stream Evolution Model Integrating Habitat and Ecosystem Benefits." *River Research and Applications* 30 (2): 135–154.

Compensatory Mitigation for Losses of Aquatic Resources. 2008. 73 *Federal Register* 70 (April 10). Department of Defense, Army Corps of Engineers, 33 C.F.R. Parts 325 and 332; and 40 C.R.F. Part 230.

DeGraff, John, and David Batker. 2011. *What's the Economy for Anyway? Why It's Time to Stop Chasing Growth and Start Pursuing Happiness.* New York: Bloomsbury Press.

Federal Interagency Floodplain Management Task Force. 1996. *Protecting Floodplain Resources: A Guidebook for Communities.* Washington, DC: Federal Emergency Management Agency and Environmental Protection Agency.

Federal Interagency Stream Restoration Working Group. 1998. *Stream Corridor Restoration: Principles, Processes, and Practices.* Washington, DC: US Government Printing Office.

Fischenich, J. Craig. 2006. "Functional Objectives for Stream Restoration." ERDC TN-EMRRP SR-52. Vicksburg, MS: Environmental Research Development Center, US Army Corps of Engineers.

Freitag, Bob, Susan Bolton, Frank Westerlund, and J. L. Clark. 2009. *Floodplain Management: A New Approach for a New Era.* Washington, DC: Island Press.

Gray, D. H., and R. Sotir. 1996. *Biotechnical and Soil Bioengineering Slope Stabilization: A Practical Guide for Erosion Control.* New York: Wiley.

Hey, R. D. 1988. "Mathematical Models of Channel Morphology." In *Modeling Geomorphological Systems*, edited by M. G. Anderson, 99–126. Chichester, UK: Wiley.

Hey, Richard D. 2007. "Natural Rivers: Mechanics, Morphology and Management." *Short Course Class Book*, 135.

Higgs, Eric. 1994. "Definitions." *Society of Ecological Restoration News*, Fall.

Knighton, David. 1998. *Fluvial Forms and Processes: A New Perspective*. New York: Wiley.

Kusler, Jon, and Larry Larson. 1993. "Beyond the Arc: A New Approach to Floodplain Management." *Environment* 35 (5):7–33.

Kusler, Jon A. 2011. *Assessing the Natural and Beneficial Functions of Floodplains: Issues and Approaches; Future Directions*. Berne, NY: Association of State Wetland Managers.

L. R. Johnston Associates. 1989. "A Status Report on the Nation's Floodplain Management Activity; An Interim Report." Washington, DC: Federal Interagency Floodplain Management Task Force, Federal Emergency Management Agency.

Lane, Emory. 1955. "The Importance of Fluvial Geomorphology in Hydraulic Engineering." *Proceedings of the American Society of Civil Engineers* 18 (745):1–13.

Leopold, Luna B., and Thomas Maddock Jr. 1953. "The Hydraulic Geometry of Stream Channels and Some Physiographic Implications." US Geological Survey Professional Paper 252. Washington, DC: US Government Printing Office.

Leopold, Luna, and M. G. Wolman. 1957. "River Channel Patterns; Braided, Meandering and Straight, Physiographic and Hydraulic Studies of Rivers." US Geological Survey Professional Paper 282-B, 39–85. Washington, DC: US Government Printing Office.

Martinez, Dennis. 1994. "Definitions." *Society for Ecological Restoration News*, Fall.

McCarley, Robert W., John J. Ingram, Bobby J. Brown, and Andrew J. Reese. 1990. *Flood Control Channel Inventory*. Misc. paper HL-90. Vicksburg, MS: US Army Engineers Waterways Experiment Station Hydraulics Laboratory.

Montgomery, D. R., and J. M. Buffington. 1998. "Channel Process, Classification, and Response." In *River Ecology and Management: Lessons from the Pacific Coast Ecoregion*, edited by R. J. Naiman and R. E. Bilby, 13–42. New York: Springer.

Montgomery, David R. 2008. "Dreams of Natural Streams." *Science* 319 (January):291–292.

Montgomery, David R., and John Buffington. 1993. "Channel Classification, Prediction of Channel Response, and Assessment of Channel Condition." Washington State Department of Natural Resources Report TFW-SH10-93-002.

National Park Service. 1996. "Floods, Floodplains and Folks: A Casebook in Managing Rivers for Multiple Uses." Washington, DC: National Park Service, Rivers, Trails and Conservation Assistance Program.

Natural Resources Conservation Service. 2007. "Stream Restoration Design," Part 654. *National Engineering Handbook*, 210-VI-NEH. Washington, DC: Natural Resources Conservation Service.

Nunnally, Nelson, and Edward Keller. 1979. "Use of Fluvial Processes to Minimize Adverse Effects of Stream Channelization." Report no. 144. Raleigh: Water Resources Institute of the University of North Carolina.

Owens Viani, Lisa. 2003. "From the Bottom Up." *Landscape Architecture* (September): 42–48.

Palmer, Margaret A., E. S. Bernhandt, J. D. Allan, P. S. Lake, G. Alexander, S. Brooks, J. Carr, S. Clayton, C. N. Dahm, J. Follstad Shah, et al. 2005. "Standards for Ecologically Successful River Restoration." *Journal of Applied Ecology* 42:208–217.

Plasencia, Doug, and Jacquelyn L. Monday. 2009. "The Need for a Resource Conservation Ethic in Flood Risk Management." Association of State Flood Plain Managers Foundation Symposium 1: Defining and Measuring Flood Risks and Floodplain Resources, Madison, WI.

Riley, A. L. 1998. *Restoring Streams in Cities.* Washington, DC: Island Press.

———. 2003. "A Primer on Stream and River Protection for the Regulator and Program Manager." Technical Reference Circular W.D. 02-#1. Oakland: California Regional Water Quality Control Board, San Francisco Bay Region.

———. 2009. "Putting a Price on Riparian Corridors as Water Treatment Facilities." Oakland: California Regional Water Quality Control Board, San Francisco Bay Region.

Roesner, Larry, and Brian Bledsoe. 2003. *Physical Effects of Wet Weather Flows on Aquatic Habitats: Present Knowledge and Research Needs.* London: Water Environment Research Foundation / IWA Publishing.

Roni, Philip, and Tim Beechie, eds. 2013. *Stream and Watershed Restoration: A Guide to Restoring Riverine Processes and Habitats.* Oxford: Wiley-Blackwell.

Rosgen, Dave L. 1994. "A Classification of Natural Rivers." *Catena* 22:169–199.

Schueler, Thomas, and Heather K. Holland. 2000. *The Practice of Watershed Protection.* Ellicott City, MD: Center for Watershed Protection.

Schueler, Thomas R. 1994. "The Importance of Imperviousness." *Watershed Protection Techniques* 1 (3):100–111.

Schumm, S. A., M. D. Harvey, and C. Watson. 1984. *Incised Channels: Morphology, Dynamics and Control.* Littleton, CO: Water Resources Publications.

Shields, F. Douglas, Ronald Copeland, Peter Klingeman, Martin Doyle, and Andrew Simon. 2008. "Stream Restoration." In *Sedimentation Engineering: Processes, Measurements, Modeling and Practice*, ASCE Manuals and Reports on Engineering Practice No. 110, edited by Marcelo H. Garcia, chap. 9. Reston, VA: American Society of Civil Engineers.

Simon, A. 1989. "A Model of Channel Response in Distributed Alluvial Channels." *Earth Surface Processes and Landforms* 14 (1):11–26.

Soar, Phillip, and Colin Thorne. 2001. "Channel Restoration Design for Meandering Rivers." Coastal and Hydraulics Laboratory ERDC/ChL CR-01-1. Washington, DC: US Army Corps of Engineers.

Society for Ecological Restoration, International Science and Policy Working Group. 2004. *The SER International Primer on Ecological Restoration.* Tucson: Society for Ecological Restoration International.

Task Force on the Natural and Beneficial Functions of the Floodplain. 2002. *The Natural and Beneficial Functions of Floodplains: Reducing Flood Losses by Protecting and Restoring the Floodplain Environment: A Report for Congress* (FEMA 409, June). Washington, DC: Federal Emergency Management Agency.

Thorne, Colin R., Richard D. Hey, and Malcolm D. Newson. 1997. *Applied Fluvial Geomorphology for River Engineering and Management.* West Sussex, UK: Wiley.

Walter, Robert C., and Dorothy J. Merritts. 2008. " Natural Streams and the Legacy of Water-Powered Mills." *Science* 319 (5861):299–304.

Williams, Phillip. 1990. "Rethinking Flood-Control Channel Design." *Civil Engineering* 60 (1):57.

Wohl, Ellen. 2005. "Compromised Rivers: Understanding Historical Human Impacts on Rivers in the Context of Restoration." *Ecology and Society* 10 (2):2.

Chapter 3

Neighborhood-Scale Restoration Projects

Many urban stream restoration projects tend to be opportunistic, reach-scale projects constructed to enhance a neighborhood or business district as opposed to projects contained in plans that set priorities for ecosystem restoration. This chapter is a selection of reach-scale projects fitting this description. They range from small-scale projects located in parks and a school ground to a large-scale housing development and city business districts. The selected projects describe a historic continuum from the early 1980s, when the concept of restoration was being discovered and defined, to the 2010s, when restoration practices and planning evolved to much greater sophistication. Each case provides a lesson in historic context, community organizing and planning, restoration design, and long-term project maintenance. Together, the cases produce common themes on how they came to be implemented and important discoveries on project designs for long-term restoration planting success. In all cases, the projects inspired more projects that followed them and therefore influenced changes in the watershed that went beyond a project's limited boundaries.

These reach-scale projects helped develop professional confidence in using restoration as a new paradigm for addressing common flooding and erosion problems in urban settings. Ultimately, they also helped developed public and political support for tackling larger, more complicated projects that followed in the San Francisco Bay Area. The cases represent the application of all the restoration schools described in chapter 2, including the use of "passive" restoration methods through the establishment of a beaver colony in a business district creek. The surprise was discovering over time the degree to which these reach-scale projects functioned as habitat.

Strawberry Creek Daylighting in Rail Yard, Berkeley, 1983

Location: Strawberry Creek Park Center, 1250 Addison Street, Berkeley, California, between Addison and Bancroft Streets, east of Bonar Street
Drainage area to project site: 2.8 square miles
Park acreage: 4 acres
Project length: 160-foot valley length, 200-foot channel length

Project History

The Berkeley Public Works Department said that the proposal to daylight Strawberry Creek as part of a park development project would kill people. The idea of digging up a creek that was safely locked underground in a culvert was one of those over-the-top infeasible ideas that Berkeley residents are famous for. The setting for this controversy is an abandoned freight rail right-of-way running north to south through the low- to moderate-income west Berkeley neighborhood. The creek flows a distance of 5.2 miles from Strawberry Canyon in the Berkeley hills to San Francisco Bay, with the project site location shown in figure 3.1.

The site consisted of a Santa Fe railroad freight yard that started in the 1800s with a trestle over Strawberry Creek, creating a nice refuge for Irish boys to have their fun (which explains why the area was called Irish Gulch). A railroad improvement in 1904 replaced the trestle with a long culvert. The railroad largely abandoned the rail in 1948, and the city purchased the site in 1974 (Wolfe 1987; Powell 1991). A large industrial building located near the tracks had been used as a bakery and was currently in use as a wood processing and woodworking shop. Also adjacent to the site was an old Wonder Bread factory in reuse as an afterschool youth help program, the Berkeley Youth Alternatives. As shown in figure 3.2, the site contained no developed use to attract public use and was therefore an open invitation to drug dealers and other unsavory characters (Schemmerling 2013).

The development of Strawberry Creek Park was part of a citywide vision to identify neighborhoods in Berkeley underserved by parks and open space and to develop parks in these areas based on needs the neighborhoods identified. Neighborhood meetings were held near potential park development sites, and public input went into the drafting of a local bond initiative, Measure Y. After the measure passed, the Berkeley Parks Department used its funds to hire two University of California, Berkeley landscape architects fresh from graduation in 1978. The design program for this 4-acre site was to provide open park space; a quiet, passive, rest areas for seniors who live in senior housing next to the park; picnic space; tennis courts; and, basketball courts to attract use by teenagers. The landscape architects, Doug Wolfe and Gary Mason, and the chair of the Berkeley Parks

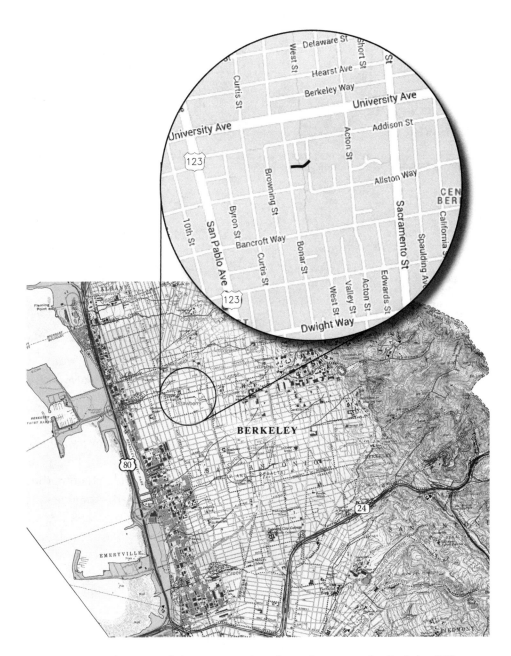

Figure 3.1 Strawberry Creek originates in Strawberry Canyon in the Berkeley Hills, flowing though the University of California campus to San Francisco Bay. The project site is located in the midportion of the watershed in a residential setting. *Credit: Lisa Kreishok.*

FIGURE 3.2 The Strawberry Creek site before the project was an abandoned railroad corridor and industrial woodworking building. *Photo credit: Gary Mason of Wolfe Mason Associates.*

Commission, Carole Schemmerling, were aware that Strawberry Creek ran under the old railroad right-of-way, with the creek open immediately upstream and downstream. A 125-foot reach was missing in the underground culvert. These creative thinkers thought that it made sense to exhume the creek, and Schemmerling came up with the term *daylighting*. Mason suggested that the park should feature the creek, but the Parks Department was insistent that Mason's suggestion was not an option. Wolfe, Mason, and Schemmerling decided to keep the idea alive by designing a park option with the creek daylighting feature (Schemmerling 2013). The project was being referred to as the SUDS project, representing the neighborhood area defined by the streets: Sacramento, University, Dwight, and San Pablo. Wolfe was quoted in a local newspaper as saying, "It was a dreadful name and pretty tough neighborhood" (Brand 1989). The first thing Wolfe and Mason did was rename the project Strawberry Creek Park as a strategy to orient decision makers and the public toward the creek (Brand 1989). Public meetings were held in the neighborhood with this alternative presented. One older man in particular expressed opposition to the idea. Schemmerling approached this individual and got him involved in the daylighting design plan, and his support developed. There would be no neighborhood opposition to this alternative plan, but how to get the City of Berkeley on board?

A veteran of local conservation issues, Mary Jeffords, a naturalist educator and a director on the East Bay Regional Parks District Board, recommended holding an additional public hearing and inviting a well-known, respected community leader to advance the daylighting concept. Her candidate was the nationally famous environmentalist, David Brower, a resident of Berkeley who played in Strawberry Creek as a boy. Schemmerling helped orchestrate an extra public hearing with all the relevant city departments present: Recreation, Parks, and Public Works. Brower took the challenge and came to the meeting, which was attended by about seventy citizens. He gave a typically rousing and commanding speech that concluded with the message that the City of Berkeley would be derelict in its duty to not embrace daylighting the creek as part of the park development (Wolfe 1987; Schemmerling 2013).

The park commissioners needed a vote to recommend the park design to the city council. Many commissioners were new and hesitant to deviate from staff recommendations opposed to daylighting. The commissioners were ultimately swayed by the public meeting and support and voted for the daylighting alternative. An important part of the city process was the support by an influential recreation director, Frank Haeg, who helped steer the concept through the city bureaucracy and develop some internal acceptance for the concept. The last political hurdle was to achieve city council approval for the daylighting park design. Frank put the park approval with daylighting on the council action calendar and strategically buried the item in a long list of consent items. The council made its vote on the consent list, and the deed was done. Subsequent complaints from a council member who did not like this strategy failed to change the outcome (Schemmerling 2013).

Measure Y contained tight budgets, and the creek restoration and park design had to innovate to stay within budget. The entire 4-acre park development and creek restoration project was accomplished at a $650,000 construction cost, with the creek restoration portion initially estimated at $85,000. Haeg was experienced with putting together funding packages and added in a federal Land and Water Conservation Fund grant to complete the budget (Schemmerling 2013). The final cost of the creek restoration component of the project was estimated at $60,000 (Mason 1993).

Project Design and Construction

The project construction drawings were completed in 1982 (City of Berkeley 1982), at a time when there were no known models to follow for designing a creek recovered from a culvert. Wolfe found a nineteenth-century map that showed the preculverting alignment, and the date on the culvert is 1904. I steered Wolfe toward the reach immediately upstream as a reference for channel restoration width and depth dimensions. This reach had gone through adjustments over the years from urbanization. The cross-sectional area had obviously widened over time, and the channel had incised below the upstream culvert. Old footbridge abutments had been undermined, but over the years, this reach had adjusted to relatively stable dimensions and had well-established vegetation growing to the edge of the active channel. By the 1980s, conditions in the built-up watershed were static, without significant changes in discharges or sediment, which suggested that the project site would not be subject to destabilizing future changes. The upstream channel reach therefore represented urban equilibrium conditions for the fore-seeable future and provided a reasonable reference. The channel slope of this upstream reach was controlled by a downstream culvert, which remained in place above the restoration project. The channel slope of the new stream was going to be controlled by both an upstream and a downstream culvert as well. The design was able to follow the historic meander dimensions, and the slope was determined by the upstream and downstream culvert invert elevations. The length of culvert removed is 125 feet long, the historic valley length was about 160 feet, and the restored channel length is about 200 feet. The sinuosity (channel length divided by valley length) ended up at about 1.25, which is typical for a creek in the mid–East Bay flatlands. Because of the upstream and downstream controls, this design was simple to construct, with few fears of instabilities. The depth from the ground surface to the culvert invert was 20 feet, so the project entailed considerable ex-cavation. The project alignment ended up following the historic channel and is shown in the design plan in figure 3.3.

The designers decided ultimately to keep most of the excavated soil on site and recycle the concrete demolished from road removal. The concrete slabs re-moved from the closure of West Street along the upstream boundary were dropped along the margins of the excavated creek, ultimately making stable channel design somewhat of a moot point. Occasional flanking of the concrete slabs and erosion by creek flows has occurred, however, illustrating the irony that there would likely have been less erosion if the concrete slabs had not been used along the channel margin.

The soil was dropped into a section of Allston Street, creating another street closure and extension of park space. Excavated soil was also reused on the site to create topographic features on a previously flat acreage. In the center of the exca-vated creek, the design added a bowl-shaped floodplain area to allow space flood flows to spread. The park design was ahead of its time in its handling of drainage by using natural swales to collecting stormwater. Significant cost savings were also realized by eliminating culverts as part of the drainage system.

This project was constructed in 1983. The removers of the early 1900s culvert

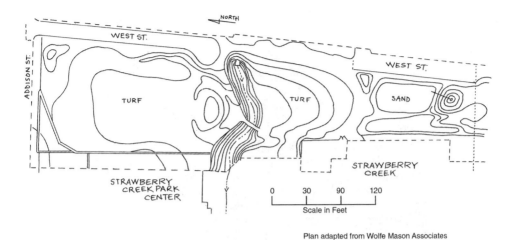

FIGURE 3.3 The Strawberry Creek design copied the upstream reach channel dimensions. *Credit: Lisa Kreishok.*

were faced with the challenge of demolishing a massive 14-foot-wide by 9.5-foot-high structure that is as thick as 4 feet in some places, representing an era of unreinforced concrete construction. The front-end loader used for excavation on the site in figure 3.4 was incapable of the demolition work, so the construction team brought in a wrecking ball to smash the culvert in. As the creek was excavated, the original location of the creek channel became apparent by following the soil darkened by ground water movement. The design layout closely followed this underground stream (fig. 3.4).

Wolfe recorded that many of his design choices were selected because of the project's budget limitations, which explains keeping most demolished materials and excavated soil on site, and replacing conventional stormwater facilities that use underground culverting with more natural surface infiltration channels (Wolfe 1987). Figure 3.5 is a photo taken at project completion.

Landscaping and Maintenance

The Strawberry Creek corridor was landscaped using a plant palette oriented toward aesthetics, color, and form in a landscape architecture design tradition. Many of the plants selected for use were California chaparral species best adapted to sunny, drier areas. The native riparian species planted included redwoods, sycamores, alders, cottonwoods, spicebush, and willow. Most of the understory and shrubs were chaparral species. An inexpensive irrigation system intended to last three years was installed, but it only lasted for the first year (Wolfe 1987; Montgomery 1993). Wolfe attributes the failure of the irrigation system to city gardeners

FIGURE 3.4 Both a wrecking ball and a front-end loader were needed to excavate the 1906 culvert at the Strawberry Creek site. *Photo credit: Gary Mason of Wolfe Mason Associates.*

who apparently inadvertently cut the shallowly buried drip lines. Because shaving costs from the project was paramount, Wolfe prescribed 18-inch planting holes filled with one-third nitrified sawdust mixed with broken-up native soil. The areas of compacted soil next to planting clusters were deliberately not cultivated, resulting in the relatively low weed populations four years later. Anticipating loss of trees by neighborhood youth, the designers built triple stake cages around trees to avoid some trampling and vandalism.

FIGURE 3.5 The Strawberry Creek construction project was completed in 1983, keeping most demolition materials on the site. *Photo credit: Gary Mason of Wolfe Mason Associates.*

Wolfe Mason Associates shielded the creek site from the north with fencing with the hope that the concrete slabs on the south bank would help protect the creek from trampling damage. The trees planted near the creek had good survival rates, but those planted back from the creek in the lawn areas did have significant losses from vandalism. Wolfe's four-year post-project report indicates that the total tree survival was 86 percent and that about 90 percent of shrubs and ground cover were surviving. He noted that the bay laurels, big-leaf maples, spice bush, and chaparral species, which included silk tassels and manzanitas, were not doing well and projected that they might not live long. Four years after project completion, he was encouraged by the canopies of the alders, willows, cottonwoods, buckeyes, redwoods, oaks, pines, and other trees. In the 1980s, the ceanothus, redbuds, and toyon were doing well (Wolfe 1987).

To address the long-term maintenance for the 4-acre park and creek restoration, the city embarked on another innovation and contracted with a youth program to take care of park and creek corridor maintenance. The city provided Berkeley Youth Alternatives (BYA) with a $25,000 annual contract for youth maintenance jobs, which went toward a program it was already supporting to help disadvantaged youth in a neighborhood with low-income project housing (Montgomery 1993).

The BYA youth, who were paid minimum wage for their work, picked up litter, raked leaves, and hoed out weeds as well as some unlucky native vegetation. The quality of maintenance improved dramatically when a city gardener got involved and trained and supervised the youth's work. The designers knew that Himalayan blackberry and other exotics would likely invade the park at some time, so they planted honeysuckle and a native grape to compete with invasive species when they arrived (Wolfe 1987).

Thirty years later, in 2013, an examination of plant survival revealed that only a few of the shrub species had survived over time and that all the chaparral species were gone. The shrub survivors were a few spicebush plants located in part sun and shade, and one hawthorn, one snowberry plant, and one dogwood grew in well-shaded locations. The ceanothus, or blue blossom, an upland species, was a beautiful bloomer for the first decade, but was relatively short lived in the damp, shady environment, and city crews removed the dead and dying shrubs in the mid-1990s. The plants that survived and thrived over time were the large trees: redwoods, sycamore, alders, cottonwoods, buckeyes, and willow, the trees that Wolfe reported as doing well in 1987. One exotic birch tree survived. Figure 3.6 illustrates the creek-side canopy provided by the native trees. (Table 4.4 in chapter 4 lists which plant species were initially planted and which survived over time.)

FIGURE 3.6 The Strawberry Creek restoration thirty years later is composed mostly of tree, not shrub, species.

My firsthand observations confirm that the park has been immaculately maintained over the years. City crews occasionally removed plants from the creek that died over time, but the creek corridor has been largely free of much maintenance outside of picking up trash and mowing the lawn. The city parks director was impressed that vandalism never became a significant issue.

Related Projects

The Strawberry Creek project generated quite a level of excitement among those with an interest in the environment. The September 1983 issue of the Bay Area Sierra Club newspaper, *The Yodeler*, dedicated its front page to an artistic rendering of the Strawberry Creek project and announced that the club was organizing an Urban Creeks Task Force. The University of California, Berkeley then awoke to the reality that the campus was developed around Strawberry Creek, a wonderful natural asset that had suffered under the university's absence of stewardship. Bob Charbonneau of the Office of Environmental Health and Safety appeared at early meetings of the Urban Creeks Council and became concerned about the mysterious discharges he noticed coming from a series of pipes entering the creek on campus. Charbonneau followed the pipes to uncover their origins and found atrocities such as chemistry department sinks discharging to the creek. He also discovered a cogeneration energy plant discharging boiling water to the creek. An initial six-month study found elevated concentrations of nitrates, fecal coliform bacteria, lead, zinc, and mercury in the creek. Cross connections between sewer lines and stormwater drains were discovered. Bob produced a creek management plan in 1987, and with a $500,000 appropriation by the campus, most of the problems were corrected by 1989 (Charbonneau 1987; Owens Viani 1999). After cleaning up Strawberry Creek, Charbonneau led the production of a "Strawberry Creek Walking Tour of Campus Natural History" (Charbonneau, Kaza, and Resh 1990) featuring the creek to induce campus awareness of the valuable resource. He was joined by the University Planning and Facilities Management office, which began to fund physical stream improvements on the campus. These efforts began in 1988 by substituting a planned concrete retaining wall along the creek with the construction of a vegetated crib wall. Failing check dams were removed, and 2,000 feet of stream channel were restored (Edlund 1988; Owens Viani 1999).

This project motivated the formation of citizen organizations to realize more daylighting projects and to also protect existing open creeks in Berkeley through policies and ordinances. Citizen creek planting and stewardship projects have been ongoing for years immediately upstream at the senior housing facility, Strawberry Creek Lodge. Shortly after the Strawberry Creek project, the city council adopted a revision to its City of Berkeley Master Plan that calls for citywide creeks preservation and restoration of creeks. A City Creeks ordinance adopted by the city council in 1989 prohibited obstructing of creek channels; required a setback of 30 feet from all channels, whether culverted or natural (thereby protecting potential daylighting options); and required a permit for creek bank projects using culverts, retaining walls, riprap, or any hard structures. The ordinance was revisited in 2004 and 2005 and received some clarifying revisions.

The East Bay Citizens for Creek Restoration, the Berkeley Citizens for Creek Restoration, and Ecocity Builders conducted numerous public creek education events and projects and pressured city officials to start the planning of daylighting Strawberry Creek in downtown Berkeley as part of downtown redevelopment as early as 1989 (Pollock 1989). This considerable public effort extended into the first decade of the 2000s with sustained involvement by the Sierra Club and other citizen creek organizations. Although the City of Berkeley considered the Strawberry Creek project an iconic representation of an innovating, green city, the downtown revival plan ultimately became defined by development interests, and the best the city could accommodate was a plan with a symbolic creek contained in a canal and fountain.

Project Lessons and Significance

The significance of this project cannot be understated. No one died, nor was any one injured. This statement may seem facetious, but the social chaos of death and injury that had been anticipated never resulted from this project and thus reset the political context for future urban stream daylighting projects in the East Bay. Today, there are five such projects in this one county. The Strawberry Creek project also introduced the now widely used term and concept of "daylighting." If we were to limit the definition of daylighting to removal of culverts, Strawberry Creek would be considered the first daylighting project in the country. Otherwise, the first daylighting project status is assigned to Napa Creek in downtown Napa, which was brought to daylight from beneath demolished housing in approximately 1977–1978. For this downtown redevelopment project, the housing was first taken down; then the large metal lid on which the housing had been sitting was removed, revealing Napa Creek flowing beneath contained within well-preserved 1930s Works Progress Administration (WPA) hand-placed rock work along the channel margins. Certainly, no matter how we determine the firsts, the public education and publicity round the Strawberry Creek project put daylighting as a concept and practice on the map nationally.

Eventually, the offices of the Urban Creeks Council (UCC), a statewide organization, and the Waterways Restoration Institute (WRI), a national organization, were located in Strawberry Creek Park Center, the renamed woodworking building at the park restored to provide office space for trades people, designers, and artists (fig. 3.7). The park and creek restoration project transformed this neighborhood from a crime-ridden area to one in which top rents could be collected for office space. The adjacent Santa Fe train station was turned into a swank restaurant, and the building later became an elementary school. Today, the park is filled with schoolchildren, senior citizens, dog walkers, and birds taking refuge from winter storms along the coast. The park received a first-place prize from the California Park and Recreation Society in 1983 and an American Society of Landscape Architects merit award for restoration and preservation in 1995. Doug Wolfe and Gary Mason, who designed the project, formed a landscape architecture firm, Wolfe Mason Associates, and applied this experience to future daylighting projects in the East Bay.

The Strawberry Creek project does not fit the definition of a restoration project. (In fact, the definition of restoration was being debated several years after this project was completed.) It is an example of an enhanced controlled channel described in chapter 2. The reuse of concrete slabs seemed an innovative and thrifty park design at the time from a historic perspective, but it is outdated because it has prevented the natural erosional and depositional processes that we want to see in creeks so as to function for water quality and habitat. Remarkably, there

FIGURE 3.7 Strawberry Creek Park now supports offices and a café in the restored industrial building located on the creek.

is no record of any hydrology or hydraulics evaluations for guiding the project design. To give the reader an idea of the scale of the flows at this project site, we can compare it with information we do have from a site 1.4 miles upstream. The drainage area to the project site, about 4 miles downstream of the headwaters, is approximately 3 square miles. Design calculations for a different location on Strawberry Creek near downtown Berkeley for about 60% of the drainage area indicates an approximate bankfull discharge of about 65 to 100 cubic feet per second (cfs), channel widths of about 17 feet, and depths of 1.3 feet, with velocities at bankfull at about 3.5 feet per second. The placement of large concrete slabs on the project channel obscured equilibrium channel dimensions, and design for stable but dynamic channel dimensions was ultimately irrelevant. The project has functioned as floodplain storage as observed during the 1986, 1995, 1996–1997, and 2005–2006 high-water years.

Glen Echo Creek Reconstruction, Oakland, 1985

Location: Between Glen Avenue and Monte Vista Streets, one block east of Piedmont Avenue, 4030 Panama Court, Oakland, California; other related projects located downstream between Monte Vista Avenue and Montell Street and along Richmond Boulevard from MacArthur Avenue to Brook Street; Glen Echo Creek flows 3 miles from Mountain View Cemetery to Lake Merritt
Drainage area to project site: 0.75 square mile
Project length: Valley length 175 feet; channel length 225 feet

Project History

In 1984, a young man ran out of his house to videotape the crime scene in his neighborhood. His camera caught the county flood control district contractor's bulldozers turning Glen Echo Creek into a cavernous ditch between Glen Avenue and Monte Vista Avenue (fig. 3.8). He was following a woman who had lain down in front of the bulldozers. She had just moved into an apartment building on Glen Avenue because it was overlooking Glen Echo Creek, and she had been awakened by the noisy bulldozers outside her window. She threw on clothes, fled her new apartment, and, not knowing what else to do, threw herself at the construction machinery to stop the destruction of the creek. Alarmed construction crews stopped their work, and in this pause the woman ran back to her apartment and started making phone calls. Community leaders appeared at the site from the Oakland Heritage Alliance, the Sierra Club, and Piedmont Avenue Neighborhood Improvement League (PANIL), among others, providing reinforcements for the tenant whose name is lost to time (Newhall 1986; Winemiller 2013).

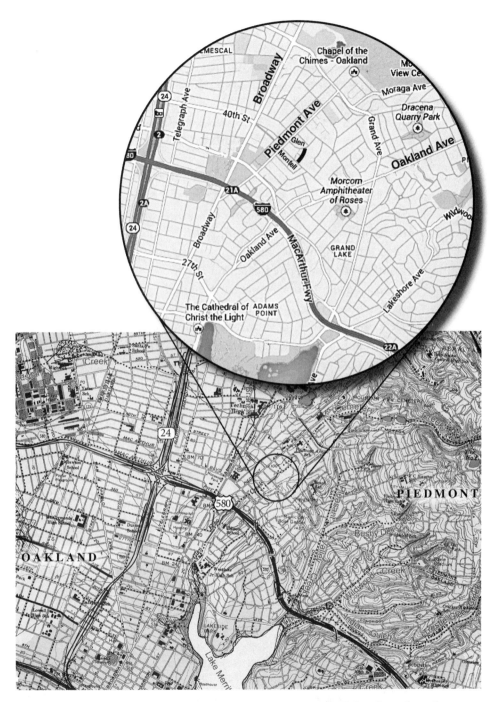

FIGURE 3.8 Glen Echo Creek drains a small, narrow watershed that flows from the Oakland Hills to Lake Merritt, which drains to San Francisco Bay. *Credit: Lisa Kreishok.*

Neighborhood leaders placed phone calls to the local city councilwoman and the Alameda County public works director. Meetings were held by the sympathetic city councilwoman, Mary Moore, for angry constituents who did not want their neighborhood creek turned into a concrete flood control channel. Their feeling was that the county flood control and water conservation district had violated its agreement that no project would occur on Glen Echo Creek without prior notification and review by the neighborhood. Ultimately, what stopped the project was the discovery that the only environmental review for the project was a ten-year-old "negative declaration" adopted by the district under the California Environmental Quality Act. This act, modeled after the National Environmental Quality Act, was supposed to provide public disclosure and review of the potential environmental impacts of proposed projects. The fact that the county had certified the environmental impact report to declare no environmental impacts for bulldozing and concreting a creek channel is a historic marker for how little regard public officials had for urban streams (Winemiller 2013).

An earlier precedent had been set for neighborhood and political interest to protect Glen Echo Creek in this location as a neighborhood amenity. Moore and her Oakland constituents had engaged with the flood control district in 1976–1977 to protect an area immediately downstream from this site, which later became part of a 750-foot long Glen Echo Creek Park. This case involved the district's condemnation of three flood-prone properties, razing of existing structures, and plans to pave over the creek and redevelop new housing. Moore and the public worked with county supervisors Bob Knox and John George to protect the creek and save the area as park space. Even a chamber of commerce–oriented city councilman, businessman Frank Ogawa, supported their cause (Feng 2013; Winemiller 2013).

The county flood control district officials realized that they had internal communication issues to resolve and were caught off guard by the incredulous and angry public response to what seemed to them a noncontroversial and routine public works project. The reasons for the project were never clear, and some concluded that it was a "make-work" project for the district as a way to expend district funds. The project reach was about 225 feet long and was located on a perennial creek between residential backyards and a senior housing facility, Satellite Homes, now called Glenwood Terrace Court. A fence in one backyard was failing and leaning, which may have precipitated the county action. Although motivations will probably never be completely known, the project that followed was something neither the engineers nor concerned citizens had ever been involved in: putting a bulldozed creek back together again as a "natural" creek.

The county flood control district selected two staff, Fred Wolin and Rick Baker, to arrive at a plan to put the creek back together again. These two had the skills to interact with the public to work cooperatively on a plan to repair the creek and became trusted ambassadors to the unfolding creek protection movement. The project objectives were to rebuild creek slopes at a 2.5-to-1 slope, assure stability in an urban environment without concrete, re-create the existing channel sinuosity at 1.28, and leave an aesthetic creek channel for neighborhood enjoyment. Ten years after the site's restoration, a community group approached the county district to request that the area be turned over to community gardening. Instead, the county staff became stewards of the creek and convinced this group that this area was not a good site for gardening and that it likewise was not good for the creek to propose the site changes needed for gardening, such as removing the riparian corridor or canopy (Feng 2013).

Project Design and Construction

What were we to do with 225 feet of creek channel that had been turned into a cavernous hole excavated out of the landscape? Some of the citizens involved in the burgeoning creek protection movement were aware of the use of gabions (wire baskets typically filled with rock and then wired together and stacked on top of each other in a staircase configuration) as an alternative to concrete and riprap bank stabilization and proposed using them as a means of rebuilding the channel. The design plan in figure 3.9 indicates the recontouring of the bulldozed channel dimensions. These citizens were also learning about erosion control from the University of California Extension Service, which had research results showing very positive results from using thick layers of straw for erosion control on road and stream slopes. The project plans called for rebuilding the creek channel to previous dimensions using gabions, but the citizens insisted on some innovations. In addition to the usual practice of filling the gabions with rock, layers of soil were also dropped in the baskets as they were constructed. The soil and rock made a planting medium while keeping the structure of the baskets sound. The gabions were stacked into the excavated bank to emulate the old stream contours so that the project did not produce an engineered stream shape. The rock- and soil-filled gabions in figure 3.10 were therefore used as building blocks to re-create the damaged banks and were planted with large tree stock such as redwoods, maples, and alders. An irrigation system was added as shown in figure 3.10, with a heavy application of straw placed on the gabions for erosion control as shown in figure 3.11. Native riparian plants were planted into the gabion soil structure shown in figure 3.12 using a design by landscape architecture firm Singer and Hodges.

Like Strawberry Creek, this early project used a controlled enhanced channel project design. Also, like the Strawberry Creek project, there is no known hydrology, hydraulics, or basis of design information prepared to inform the project design. This project purpose was for a stream bank stabilization project for a

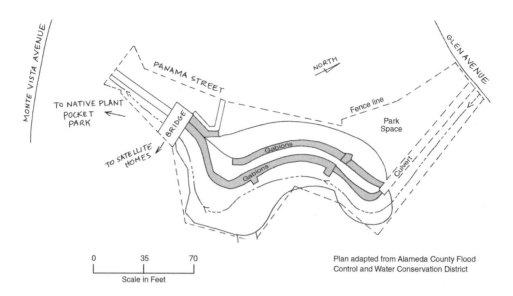

FIGURE 3.9 The Glen Echo Creek design objective was to recover the creek from bulldozing damage. *Credit: Lisa Kreishok.*

FIGURE 3.10 Glen Echo Creek was reconstructed using gabions filled with dirt and an irrigation system to support revegetation.

FIGURE 3.11 Straw erosion control was used to stabilize the final grading.

FIGURE 3.12 Plants were set into the gabion-soil medium.

corridor that is 60 feet between two terrace tops, which re-creates the same dimensions of the creek that had been damaged. The active channel width of 12 feet was retained as a channel width, and the depth is about 1 foot. The substrate is a mix of gravels, cobbles, and boulders, an inherently stable channel type. A bridge and sewer line required a grade-control structure constructed by placing a gabion below the level of the creek bed to hold the existing grade. The project costs were $250,000, which the flood control district readily accepted because the costs were greatly lower than the initially planned concrete channel.

The neighborhood association decided that given that this reach was situated between backyards and a senior housing facility on the opposite bank, it was appropriate to restrict public use to the site. A number of creek advocates in the neighborhood association considered that to be a positive decision because they believed that this reach of creek could be set aside and protected for wildlife habitat and that it was good to have a certain amount of restrictions on public use to help wildlife in an urban setting. As a result and with the support of the creek community, a chain-link fence was put up around the perimeter of the site (Wilson 1992).

Landscaping and Maintenance

The landscaping plan used a "palette" of eighteen different California native species. Of these species, five were chaparral species for drier, nonriparian environments and the rest were native trees: redwood, bay, alder, maple, a native cherry, and riparian shrub species. An irrigation system with bubblers was installed, and a pre-emergent weed killer was sprayed on the site before plant installation. A 2012 visit to the site with the landscaping plan and plant list indicated that three species of trees survived and thrived: redwood, alder, and maple. A wild grapevine and only one shrub species survived, *Myrica californica*. Elms, fan palms, equisetum, and cottonwoods have volunteered into the site. The only native plant community remaining consisted of the large-canopy trees. An extensive ground cover of non-native Algerian ivy covers much of the site. Photos of the site over time indicate that the rebuilt bank slope on an east-facing slope was open and exposed to the sun for more than five years. Fred Wolin reported in 1989 that there was an 85 percent survival rate of plants in the first four years, although no records exist as to a survival rate by species (Wolin 1993). Figure 3.13 represents 1988 conditions, and figure 3.14 illustrates how the site looked in 2013 crossing the bridge over the creek leading to the Glenwood Terrace Court Senior apartment building. The site today is dominated by redwoods and cottonwoods. My field visits in the early 1990s indicated a lush growth of the redwoods in the gabions, the disappearance of ferns and other understory plants, and a progressively intense invasion of fennel into the sunny, exposed areas.

FIGURE 3.13 The reconstructed Glen Echo Creek channel was planted with a range of native riparian plants, including ferns, shrubs, and trees. This photo was taken in 1988, three years after installation.

FIGURE 3.14 Glen Echo Creek evolved over time from a complex plant environment to one dominated by large canopy trees. This photo shows conditions in 2013.

Wolin monitored the site for at least five years and discovered that redwood trees planted with 1 gallon and 5 gallon stock were of similar size after four years. He observed a returning pair of mallards that use the creek as a summer rearing area for their young, and he counted fourteen species of birds using the site.

An informal group of advocates for the creek organized occasional maintenance projects on the site under the auspices of PANIL. The friends of the creek group had a key to the lock on the creek refuge's fence, which enabled an occasional volunteer maintenance work to pull out fennel (Wilson 1992). Ultimately, the site's maintenance is the responsibility of the district, which has sprayed elm sprouts and other invasive plants (Feng 2013). PANIL members observed that when Wolin and Baker were involved in overseeing maintenance and monitoring the creek, the project had the appearance that it was thriving. This project reach has never had a formal maintenance agreement with the City of Oakland or citizen organizations, as occurred later with downstream reaches of Glen Echo Creek. Eventually, the district flood control facilities maintenance crews received clear assignments to maintain the creek project. Herbicides were freely used, and many of the plantings were cut down with gasoline-powered string trimmers. The few species of plants remaining on the site can be attributed in part to the resiliency of trees to survive the inadequately trained flood control maintenance crews (Winemiller 2013).

The maintenance of the downstream Glen Echo Creek Park area was negotiated in 1997–1998 when the community advocated that the county's vacant lots immediately downstream of the creek reconstruction project be developed as a park. The county flood control district stated that it was not in the "parks business" and agreed only to retain maintenance responsibilities on the creek channel itself. The district and the city negotiated an agreement referred to as a "license" that maintains the site as county district property, but the city assumes responsibility for the public access and park use and maintenance. Part of this license agreement is that the neighborhood will commit to maintenance of the park. The city picks up trash, but local volunteers organize park and creek maintenance.

This example is an interesting model for creek maintenance, and participants report ongoing success in stable neighborhoods. The City of Oakland formalized a process for local citizen organizations to adopt open spaces, parks, and waterways. At this time, a formal Friends of Glen Echo Creek group was organized and created the maintenance partnership with the City. Arleen Feng is recognized as a neighborhood hero for her long-term commitment to organizing this volunteer involvement on Glen Echo Creek. Her role has been more than the official weed controller and includes interacting with the public to prevent inappropriate vegetation removal and intercepting ideas for land conversions that would compromise the creek environment. The downstream section of Glen Echo Creek has its own greatly appreciated steward, Joe Trapp, who has developed a nursery of native plants by collecting native seed and cuttings to improve the Oak Glen Park area with neighborhood plantings. A most unfortunate tragedy occurred before the city stewardship program began when city crews, uninformed of constructive citizen efforts at stewardship, "cleaned up" Joe's nursery and set back neighborhood efforts to vegetate the corridor. This incident reinforces the issue that the education and training of government maintenance crews are critical components of stream stewardship. Since then, regular neighborhood planting and invasive removal projects have taken place in this neighborhood.

Related Projects

The interest in highlighting the Glen Echo Creek corridor as an amenity dates to the early 1900s and developers interested in creating a memorable, beautiful housing development in Oakland. Neighborhood interest and ongoing volunteer activities started anew in the early 1980s, representing one of the earliest and most sustained neighborhood efforts in the San Francisco Bay Area to regreen the creek, foster native riparian plans, protect remaining sections of the creek from development impacts, and, most importantly, foster a budding new urban creeks movement.

OAK GLEN PARK, 1911–PRESENT

At the beginning of the twentieth century, Glen Echo Creek appeared in "city beautiful movement" plans in which planners and architects envisioned green-belts winding through the City of Oakland. The city did not take the initiative to implement this plan but the plan inspired the developers, Frank and Wickham Havens, to follow this concept. The Havens developed the Richmond Boulevard and Brookside Street neighborhood area after the purchase of the Oak Glen Park area in 1905. They protected a 1,500-foot reach of creek and riparian forest through the center of Richmond Boulevard, and the lots were laid out following the meander of the creek. A beautiful stone bridge and pergola were added in 1911. Majestic coastal live oaks follow the Oak Glen Park creek corridor (Newhall 1986; Starr 1989). This unlikely oasis is set in a dense urban environment known for high crime statistics, bordered by car dealerships and major transportation arteries running through Oakland: MacArthur Avenue and Interstate 580. A creek improvement project in this neighborhood in 1982 helped inspire the formation of the Urban Creeks Council as well as the state-administered grant program established by this organization to support urban stream restoration throughout California (Pollock 1989).

PLANTED CREEKSCAPE AND BRIDGET'S WPA-STYLE WALL, 1983–1984

The creek improvement project on Richmond Boulevard was implemented in 1982 on a private residential property by the owners, Rebecca Walden and Bridget Brewer. The project addressed a failed creek retaining wall and used plants and hand-laid rock to rebuild the failed walls for 50 feet. Both women were influenced by their experiences with the California Native Plant Society and their relationships with a new generation of landscape designers who were installing gardens for residences and public spaces using native California plants rather than the standard practice of using commonly available ornamental nonnative nursery plants. (Rebecca Walden is an attorney and carried a card with her self-assigned title, "Attorney at Lawn.") This movement received a big boost from the 1976–1977 drought in California, which created a demand for drought-tolerant, low-water-using landscapes, for which native plants are well adapted. This period also resulted in a proliferation of demonstration landscapes using native plants, increasing public awareness and demand for California native plants and creating native-style landscapes.

The WPA-style rock wall along Glen Echo Creek on Rebecca and Bridget's property originally constructed in 1925 failed in the flood of 1982. Bridget and two helpers rebuilt the WPA-style wall by hand and interplanted the spaces between the rocks with native cuttings such as dogwood. They completed the project in four weeks for a total cost of $2,500. The funding came from a pilot urban stream project started by the California Department of Water Resources (DWR) using federal Water Resources Council funds, which at that time were distributed to state water resources centers to encourage innovative water management projects and research. Theirs was truly a native plant gardening project next to a creek, with the innovation that the walls were interplanted with cuttings of native stock and that native riparian plants were added to the stream corridor. During this time, urban creek restoration was considered an offshoot of the urban beautification movement using native plantings. The people involved in this project helped organize a neighborhood-wide effort to rid the creek corridor of invasive exotics and worked with the City of Oakland to place large tree trunks cut from city tree-removal projects along the creek corridor to prevent parking and dumping next to the creek.

An important feature of Bridget's wall was that it was located next to residences with historic significance and so appeared as part of Oakland Heritage Alliance walking tours. The great appeal of this project spread through these tours, and the urban streams movement realized that demonstration projects and tours would be central to its strategy to develop public education on the options for managing urban waterways. "Seeing is believing" became the motto.

Monte Vista Pocket Park, 1997–1998

In 1997–1998, another member of the landscaping with natives movement, Michael Thilgen, worked with a group of friends and planted native vegetation on the banks of Glen Echo Creek in the reach below the county reconstruction project located between Panama Street and Monte Vista Avenue. This project was also an early example of landscaping with native plants in creek environments with an emphasis on calling public attention to their aesthetic qualities. The neighborhood association reports feeling unprepared to handle the maintenance for the area and did not have an irrigation system to work with to achieve plant survival. The pocket park nonetheless remains an attractive amenity with a few good surviving species such as oak, coyote bush, toyon, buckeyes, myrica, mahonia, and willow (Winemiller 2013).

Glen Echo Creek Park, 2002

Michael Thilgen, a resident of Oakland and a community-minded landscape architect, advocated for the development of the three-lot open space downstream of Monte Vista, which had been protected from development in the 1970s, into a more formal public-use park. Although the creek had been saved from culverting, this three-eighths of an acre open space had been largely abandoned by the district and had devolved into a drug-dealing haven. The neighborhood ultimately succeeded in bringing public focus to the site and developing it as a pocket park.

The next episode on Glen Echo Creek raised the question, Can riparian plants move from beautiful form to equally beautiful function? By 2001–2002, the City of Oakland had hired new staff with an environmental commitment to foster watershed management and creek protection. The staff had good exposure to the newly evolving alternatives for creek management, including soil bioengineering. A fallen oak tree on the largely abandoned site created a test of how this section of creek would be managed. The competing concepts were conventional engineering, a preference of the county flood district, and the new soil bioengineering planting approach, which was favored by the city staff. This issue aligns the City of Oakland staff, the neighborhood, and creek groups against the district, which wanted to see conventional engineering with proposals that evolved from concrete, large logs, or boulder-lined channels with plants inserted into tubes in the boulders. The San Francisco Bay Water Resources Control Board, the regional water pollution control agency, became the ultimate arbitrator and informed the district that its conventional proposals would not receive the required permit to work along the creek.

A demonstration soil bioengineering project was installed by Jorgen Blomberg, a landscape architect who had just started working for the engineering firm Philip

Williams Associates. The participation of an engineering firm that was starting to use soil bioengineering helped the flood control district make the leap to the new nonstructural technology for stabilizing creek corridors. The soil bioengineering approach was used, and project performance has been good by everyone's standards. The creek steward for this area reports that the willow growth has endured, kept the banks stable, and survived through time when other shrubby plants were not sustainable. Like the upstream gabion reach, many container shrubs and herbaceous plants from the 1989 planting project at Monte Vista have not survived to provide functions of shade, bank stabilization, or habitat (Estes 2013; Feng 2013). Blomberg benefited from his involvement in smaller neighborhood-scale projects to develop his expertise in restoration and soil bioengineering, which he later applied to large regional-scale projects.

OAKLAND CREEKSIDE PROPERTY ACQUISITION, 2003
The saga of Glen Echo Creek ends appropriately with a citizen's tree hug-in on property at the downstream end of the Oak Glen Park area. The tree hug-in was organized for early morning hours where Richmond Boulevard meets Brookside Street to intercept logging crews hired by a creek-side property owner. The property owner of this very steep site had inappropriate development plans given the severe site constraints and had started to remove trees to better position himself for development approvals. The trees were saved, and after sustained protests over the years by the neighborhood residents, the city convinced the county to buy the lot in 2003. As progress is made to protect and restore the creek, this public open space will anchor the Glen Echo Creek necklace of parks through the middle of Oakland. The city attributes the successes along the Glen Echo Creek corridor to the citizenry and understands that the most effective strategy for local government to assuring stewardship of the creek corridor was to form relationships with the Friends of Glen Echo Creek (Estes 2013; Winemiller 2013).

Project Lessons and Significance

In concert with other nearby stream restoration and stewardship activities on Glen Echo Creek, this project contributes a historical understanding of the incremental, unfolding practices of urban stream restoration. It is also a valuable reach to evaluate because we have the original plant installation plans and project records for the 1985 project. These records, combined with the original plans and long-term records of seven other projects in the East Bay, develop a record on riparian planting results recorded over time. This subject is covered in greater detail in chapter 4, which contains a table of plant survival over time for a number of projects.

The Glen Echo Creek story records the historic evolution of the urban streams movement. It progresses first from public efforts to protect open space along the highly urbanized creek environment in the latter 1970s to the next stage in which some people became involved in the 1970s to 1980s in the landscaping with native plants movement. This stage began with planting California native riparian plants along the creek and removing nonnative, exotic, and invasive plants. The Glen Echo Creek reconstruction in the mid-1980s makes the progression to an actual physical rebuilding of a damaged but still "naturalized" stream channel. The Glen Echo Creek story then progresses from this site and time to the development of a creek-side pocket park with the use of soil bioengineering in the late 1990s to both restore a functioning, dynamic creek channel and address serious channel instabilities.

The Glen Echo Creek gabion reconstruction at Glen Avenue demonstrated the physical restoration of a creek environment without concrete or boulders. The soil bioengineering project downstream of that reconstruction demonstrated that physical reconstruction of creek environments could occur with plants and need not be dependent on engineering structures such as gabions, boulders, crib walls, or retaining walls. We also learned that a project implemented for engineering purposes could simultaneously recover some ecological values. This case and the related projects also illustrate the progression from thinking of creek "restoration" as gardening with native plants to using plants for the important functions of shade, water quality, and stream bank stability. Finally, a new generation of public employees in cities and counties, an engaged public, and the evolution of regulatory participation combined to change how this urban stream has been managed. My travels throughout California and the United States make this case representative of thousands of others occurring at this time.

The final postscript to the projects along Glen Echo Creek has an unexpected lesson. The beautiful and native creek-side garden that tours featured on Richmond Boulevard at Bridget's WPA wall is no longer there, having succumbed to neglect and weeds. The wall is composed of Algerian ivy, colonized from a nearby lawn. The trees—redwood, buckeye, and oak—are all that remain of the lovely native garden. The property was sold in the 1990s; the creek-side residents in this block have limited knowledge and resources to maintain the project, and the new landlord does not maintain the property in the acquired enhanced condition. The renters live there because of the creek, but the previous property owners' efforts to maintain native riparian plantings at this site are no longer. Neighborhood efforts to prevent dumping in the public park in the Richmond Boulevard area have succeeded, though, and their tree-planting projects leave a legacy. They are still fighting against invasive Himalayan blackberry and Algerian ivy.

The so-called conventional wisdom known as the tragedy of the commons

(Hardin 1968), in which public property or resources are neglected from lack of private "ownership" and privately owned property is better managed, has now been stood on its head. In the Glen Echo Creek case, the public spaces are sustained with long-term management and maintenance, while it is the project on private property that loses its environmental values. The citizen volunteer–city maintenance regime supported by the City of Oakland is working well in the long term in Glen Echo Creek Park, Monte Vista, and Oak Glen Park.

Blackberry Creek, Daylighting in Thousand Oaks Elementary School Yard, Berkeley, 1995

Location: Thousand Oaks School playground between Solano Avenue (to the south) and Tacoma Avenue (to the north) and Colusa Avenue (to the east) and Ensenada (west), Berkeley, California
Drainage area to the project site: 0.3 square mile, drains to subterranean Marin Creek
Project length: 200-foot valley length, 240-foot restoration channel
Park acreage: 0.6 acre

Project History

The Blackberry Creek daylighting project (fig. 3.15) in an elementary school yard is ultimately a PTA mom project (the pre-soccer-mom era). Mimi Roberts, who had school-age children and was active in the school's parent-teachers' association (PTA), had the success of the Strawberry Creek daylighting project as wind in her sails and the backing of a now-maturing urban creeks movement. A 1947 aerial photo of the Thousand Oaks School shows Blackberry Creek winding through the length of the property, east to west, with a well-vegetated riparian corridor of mature trees. A student from this school who played in this creek in the early 1930s became a well-respected San Francisco Bay Area leader in innovative, experience-based environmental education starting in the early 1950s with the Audubon Society. This leader, Mary Jeffords, was the elected representative to the East Bay Regional Park District Board of Directors from 1973 to 1981 and the first woman to be chair of the board. She was much loved by her constituents, although her forward vision and uncompromising representation of idealistic constituents from her East Bay district was known to occasionally create angst with park district staff and managers. Jeffords never recovered from what in her mind was the ultimate atrocity of the filling of Blackberry Creek in her elementary school yard in the 1960s. She constantly and relentlessly reminded her friend Carole Schemmerling, who sat on the Urban Creeks Council board of directors, that the council

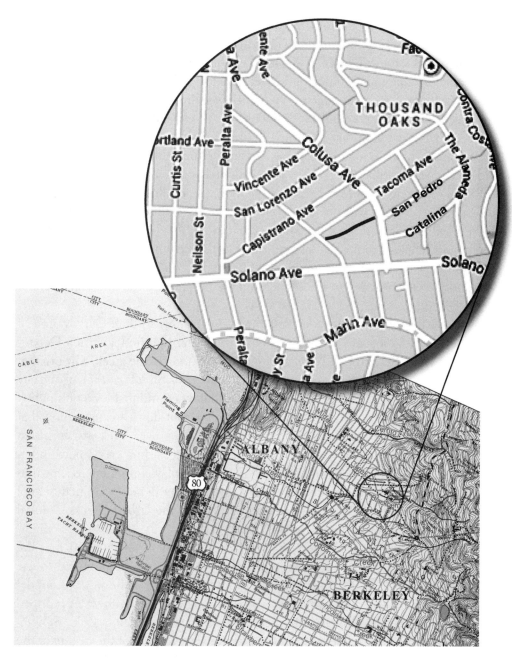

FIGURE 3.15 The Blackberry Creek project is located in an elementary school yard in Berkeley adjacent to a commercial district. *Credit: Lisa Kreishok.*

had the responsibility and duty to bring the creek back to the schoolyard. Schemmerling shared this need with Roberts, a friend met through a local political club. Mimi had children who were going to reach the age to attend this school, and she became the champion of moving Mary's agenda and legacy.

Blackberry Creek was largely an opportunistic project tied to a redevelopment project scheduled for an old 1913 school structure to address structural deficiencies to withstand earthquakes. The 1989 Loma Prieta earthquake led to seismic evaluations of all Berkeley's schools, a reaction and obligation of many school districts in California. As a result, a $158 million bond measure was adopted by the Berkeley citizens in 1992 to make their schools structurally safe from earthquakes. Because new plans were being developed for the Thousand Oaks School site, the discussion within the school district and the PTA turned to the possibility of including changes to the school grounds as well. The school district hired a team of architects and landscape architects—Stanley Saitowitz, Rosemary Muller, and Walter Hood—to redesign the school and the property (Thousand Oaks School 1994).

In the meantime, Roberts exploited this opportunity to draft Gary Mason of Wolfe Mason Associates in 1992 to help write a grant to the California Department of Water Resources (DWR) to fund the restoration of the creek. The original vision was to restore the creek from Colusa to Ensenada Streets, a reach of one city block long, but the DWR grants program could not commit that level of funding. The daylighting focused at this point on the 200-foot downstream section that occupied a neighborhood park referred locally as "the grove." This area consisted of large trees that were on or near the bank of the pre-existing creek, now underground, as well as bare dirt and unsafe and outdated play equipment as shown in figure 3.16 The DWR awarded a $144,000 grant to the Berkeley Unified School District and the PTA, the lead grant applicants, and the Urban Creeks Council which served as the community cosponsor.

The school district architect, Saitowitz, had originally determined that the creek culvert was too far underground to exhume. Instead, he proposed to use the Italian Villa D'Este, a garden with fountains outside of Rome, as a model and run a faux creek in a symbolic channel through the site with water pumped up from the creek (Berkeley Architectural Heritage Association 1996; Schemmerling 2013). Urban Creeks Council members were aware of the history of the well-intended faux creek designed at the Claremont Avenue, Oakland, California Department of Motor Vehicle grounds in the 1970s. This project used the strategy of pumping water up from buried Temescal Creek to provide a garden-like environment with a faux creek flowing through it. This experiment taught us that these projects would not be sustainable because the pumps frequently broke down and

FIGURE 3.16 The Blackberry Creek site before the daylighting project contained a neglected, outdated play area.

neither the city nor the Department of Motor Vehicles could carry out high-level maintenance. Ultimately, the expense was too great to maintain this system.

Roberts, Wolfe, and Schemmerling kept the focus on the phase one daylighting project, and Mason had numerous meetings with the Thousand Oaks Neighborhood Association and city officials to advocate for the project. The neighborhood association was predominantly in favor of the daylighting, but three very vocal people opposed the project because they did not want to lose the children's play equipment. Ultimately, the daylighting project did proceed, and it was completed in 1995, before the old school was razed and rebuilt. A new tot lot was constructed three years later, which put to rest neighbors' complaints about the creek project. The school district architect was replaced with a new firm that embraced the daylighting feature, Muller and Caulfield, that designed the school that is currently on the site. School reconstruction began in 1998. The site is designed so that a bridge spans the still culverted creek in the upstream eastern portion, thereby preserving future options for the second phase of a Blackberry Creek daylighting project. Walter Hood, a professor at the University of California, Berkeley, designed the play area (Berkeley Architectural Heritage Association 1996, 2000; Schemmerling 2013).

Statewide and National Context

The larger context for this 1995 project is important. A critical development was the maturation of the urban creeks movement, which included a more evolved definition of environmental restoration, accumulated project design and construction experiences, and accumulated significant skills from exposure to stream restoration projects starting to occur both statewide and nationally. Equally important was the development of a stable funding source from the State of California designed to serve the grassroots nature of the urban streams community.

By the time the Blackberry Creek project was being planned, the Urban Creeks Council had organized a statewide network of urban creeks organizations, and this political organization resulted in getting a dedicated source of state funding for urban streams restoration through state legislation. This extraordinary effort, which began in 1983, eventually helped write and shepherd a bill through the state legislature and persuade a conservative Republican governor, George Deukmejian, who up to this time had a consistent record of opposing environmental measures, to sign the bill. The governor's initial veto of the bill, in 1983, killed the urban creeks legislation. In 1984, the urban creek advocates initiated an important change in strategy at the urging of the bill's author and solicited a conservative Republican legislator, Eric Seastrand, as a new author to the bill, replacing the previous lead author, Tom Bates, a liberal Democrat of Berkeley. Seastrand's constituents included a water district wanting to improve the Carmel River and a business community that had been enhancing and restoring San Luis Obispo Creek in a downtown business district. Republican State Senator Milton Marks from environmentally oriented Marin County became the bill's cosponsor in the state senate, and letters from Republican and moderate Democrat state legislators seeking support were mailed to the governor's office. This move established that the urban creeks movement knew no particular party identity and that Republican and Democrats alike had stream issues to solve in their districts. Citizens involved in this effort also learned that they could not negotiate state capital political dynamics on their own and passed the hat to fund political assistance from Jerry Meral, now an environmental advocate in the nonprofit sector who had previously supported the earliest state-sponsored urban stream restoration efforts dating to 1981 as deputy director of the DWR. With legislation that now represented the political right as well as the political left with the coauthors Seastrand and Bates, Deukmejian signed the Urban Creeks Restoration and Flood Control Act of 1984 in 1985, after vetoing 85 percent of bills authored by Democrats that year. This legislation brought in a pioneering restoration program located in an agency mostly noted as a water conservation and supply agency, the California DWR. The 1985–1986 fiscal budget contained the first state funding for a statewide urban creeks grant program (Schemmerling 1984).

The Urban Creeks Restoration and Flood Control Act of 1984, acquired political legs because it recognized that restoration projects were a new, multi-objective strategy to address common urban stream erosion and flood hazards with practical but environmentally friendly solutions. To be eligible to receive funding from the grant program, applicants have to show that they are addressing a flood or erosion hazard. A unique measure in the act requires an appropriate community, creek, or environmental organization to be an applicant along with an appropriate local government agency such as a public works department. Therefore, no project can receive funding that the local community groups cannot support (such as a conventional channelization or riprap project); likewise, no project can be funded without local agency sponsorship, which avoids antagonizing and polarizing public works agencies still accustomed to single-purpose conventional engineering projects. This provision became responsible for new alliances forming among parties that may have had antagonistic histories. There is nothing like a new source of funds to pull disparate parties together. I had the good fortune of administering this program for DWR in its earliest years until 1990, and it has retained its grassroots culture today.

The evolving experience of the project designers for the Blackberry Creek project was tied to the growth of the stream restoration movement both statewide and nationally. Networking among design professionals and citizen groups through the newly formed national Coalition to Restore Urban Waters in 1993 exposed many of us to developments in project development and design in places such as Chicago, New York, Atlanta, and New Orleans. The continuation of small, local East Bay restoration projects at sites such as Seminary, Courtland, and Sausal Creeks in Oakland gave us increasing confidence to design equilibrium channel dimensions and acquire experience constructing soil bioengineering systems. The urban creeks community sponsored hands-on workshops with pioneers in the soil bioengineering field, including Robbin Sotir and Andrew Leiser, to both teach us and expose our colleagues in consulting firms to new techniques of stream channel stabilization (Gray and Leiser 1982). The Society of Ecological Restoration brought wider attention to integrating fluvial geomorphology, plant ecology, and animal ecology into project design and stressing design for ecological functions and stream dynamics. My own professional development was greatly influenced by participating side-by-side with colleagues in Atlanta, Minnesota, and Chicago as well as practitioners in California whom I met through the DWR Urban Streams Restoration Program. This program also took on an important role as a statewide networking and educational center because participating in project development and design and observing the results became the ultimate instructing experience.

According to Oscar Wilde, "Experience is the name we assign to all our mistakes." At the time I was preparing a creek restoration design report for the project at Blackberry Creek, I was part of an effort to resolve the comedy of errors unfolding at an underfunded, "informal" daylighting project. This project involved at least three uncoordinated citizen organizations that had inadequate design plans to daylight Codornices Creek at Ninth Street next to the new site for the Body Time Company in Berkeley. The lessons learned here carried us to better project planning, permitting, budgeting, design, and design review. Inadequate communication and design produced a daylighted cross-sectional area for a bankfull channel of 12 square feet instead of 30 square feet. One organization viewed daylighting as just digging up the creek without a design informed by fluvial geomorphology or hydraulic considerations and allowing the stream to make the ongoing needed adjustments. The concept of passive restoration approaches for self-forming channels was encouraged by university professors, so this strategy had followers. A naive attempt to save the unsustainable cross section with soil bioengineering failed catastrophically, resulting in large amounts of bank, slope, and willow cuttings carried down to San Francisco Bay in the next winter's rainy season. Unguarded excavation equipment left on site without fencing or security hires attracted self-described "punks" from a nearby music club who vandalized equipment. You name it, we all made the mistakes.

Ultimately, a new design, excavation, and planting had to occur. The City of Albany provided a small emergency budget to address the problems. Many of us referred to this site years after as our "outdoor classroom of mistakes" and therefore one of the most valuable sites we were ever involved with. Later, a long-term, sustained volunteer planting project was conducted at this site by a local organization, Urban Ecology, which turned this area into a valued pocket park.

In fact, learning from our mistakes was enshrined in a tradition. At the Coalition to Restore Urban Waters' annual national conferences, a scheduled open session was held in which stream restoration practitioners and organizers could stand up and confess their project mistakes to the applause of the audience. The worse the mistake, the louder the clapping, encouraging sharing without judgment and helping one another avoid similar mistakes in the future.

Project Design and Construction

The design shown in figure 3.17 used a combination of information from upstream reference sites, multiple regression equations for determining flood frequency in the Bay Area developed by the US Geological Survey (USGS), Manning's equation to determine culvert capacity, a pipe-head loss equation, regional hydraulic geometry using a Bay Area regional curve, and the rational method as a check on

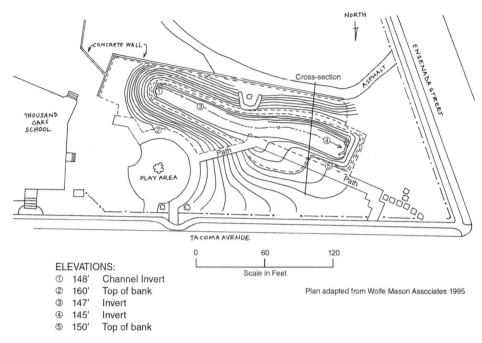

ELEVATIONS:
① 148' Channel Invert
② 160' Top of bank
③ 147' Invert
④ 145' Invert
⑤ 150' Top of bank

Plan adapted from Wolfe Mason Associates 1995

FIGURE 3.17 The Blackberry Creek restoration design followed the creek's historic, meandering path and fit the channel between existing large trees. *Credit: Lisa Kreishok.*

the hydrology (Riley 1994; Szumski 1995). A 1947 aerial photo informed the planform for the creek, but there was no upstream or downstream reference condition to gauge a restoration width and depth. Tree cover in the photo made it difficult to discern the sinuosity.

Table 3.1 displays the different values that we computed for the hydrology and channel dimensions to inform the design. We went first to the Leopold regional curve developed for the San Francisco Bay Area (Dunne and Leopold 1978) to provide our first estimate for channel dimensions based on the drainage area of 0.3 square mile: about 11 square feet cross-sectional channel area. Because the neighborhood upstream was located on a steeper 10 percent valley slope, and the restoration site was on a 4 percent valley slope, the upstream channel widths and depths were not going to be able to inform downstream width-to-depth ratios for a restored stream, but they would be useful in finding indications of cross-sectional areas. I started knocking on front doors in the neighborhood upstream where the creek surfaces. I explained that I was hoping to see Blackberry Creek in the property owners' backyards and was relieved to get friendly and warm invitations to the yards. The property owners often had vivid memories of the creek's high-water elevations in the 1982 and 1986 floods, and one owner pointed out the 1955–1956 flood line on his back steps. I saw different creek widths and depths based on how the creek was constrained by gardening walls and landscaping features.

One property owner told me that the variety of creek dimensions was in part due to the stream being filled in after the great Berkeley fire of 1923. Digging around in backyards in this neighborhood brought the discovery of mounds of old pots, boots, pieces of furniture, and other relics. Neighbors trading information developed the knowledge that the debris they held in common in their backyards came from the discarding of fire ruins from the older neighborhood above them. This information was important because restorationists are trained to expect that urban streams are generally wider than historic conditions. This lesson about urban fill informed a future project on Codornices Creek at Fifth Street in Albany, which also had significantly smaller channel cross sections than regional averages. (A tip we learned was to add debris removal and off-site hauling to a restoration budget.) Despite the variances in channels widths and depths, the cross-sectional areas of many of the least modified channels averaged 9 to 10 square feet, close to the regional curve value. Using this reference information and regional curve data, the design dimensions selected for the bankfull channel were 8 feet wide and 1.5 feet deep. The width-to-depth ratios were informed by regional values on channel widths and depths using the regional curve shown in figure 4.1 in chapter 4.

The hydrology was estimated using a number of sources from regional regressive analysis and in the field flow measurements. US Geological Survey (USGS) multiple regression analysis is based on hydrologic records from forty gage stations in the bay area, precipitation, and drainage areas and uses a degree of urbanization coefficient (Rantz 1971). Calculations using USGS multiple regression provided a value of 16 cfs for the two-year recurrence interval (RI) flow. The Leopold regional curve value for bankfull discharge based on drainage areas in the Bay Area was a close 15 cfs. The multiple regression five-year RI discharge was 40 cfs, the ten-year RI was 60 cfs, and the fifty-year RI was 160 cfs. The one-hundred-year RI was extrapolated on a flood frequency curve at 220 cfs. On February 21, 1994, I measured a full channel flow in the backyard of a cooperative resident whose property had one of the best reference conditions and recorded a velocity of 1 foot per second with a float and stopwatch. This measurement helped support the reasonable magnitude of the regression and regional curve value for bankfull discharge estimates. Summer visits to the best reference site upstream indicated a low flow channel of about 2 feet wide and 0.5 foot deep. The project engineer computed a check on the hydrology using a rational method calculation, inputting a runoff coefficient and rainfall intensity estimates in use by Alameda County Public Works Department. This value collaborated very closely with others estimated, with the 1-in-50-year RI computed at 175 cfs and the 1-in-100-year RI at 220 cfs. We drew an estimated elevation for the fifty-year flood on the design cross section using a

TABLE 3.1.

Blackberry Creek Design

Hydrology	1.5- to 2-Year Recurrence Interval Discharge (cfs)	10-Year Recurrence Interval Discharge (cfs)	50-Year Recurrence Interval Discharge (cfs)	100-Year Recurrence Interval Discharge (cfs)
San Francisco Bay Area regional curve for 0.3 mi² (Dunne and Leopold 1978)	15			
Regional multiple regression (Rantz 1971)	16	60	160	220
Alameda County rational method			175	220
Field measurement for bankfull discharge	10			

Channel Dimensions	Cross-sectional Area (ft²)	Width (ft)	Depth (ft)
Reference reaches	9–10		
San Francisco Bay regional curve for 0.3 mi²	13	11	1.2
Project design	12	8	1.5
1996 post-project adjustment survey		5.5–7	0.43–1.15
1999 survey		6	
2005 survey		5–6	1.0–1.2

reasonable guesstimate by applying Leopold's hydraulic geometry relations from western streams based on dimensionless rating curves indicating that, on average, fifty-year RI flood elevations tend to be about twice bankfull elevations (Leopold 1994). This estimate was collaborated by the neighborhood observed flood elevations from the historic high flows. The one-hundred-year flood at best would achieve an elevation of 3 to 4 feet depth of flows, leaving an additional remaining 7 feet of channel depth above this flow level to absorb any greater catastrophic flood flows. Plenty of freeboard also remained for flows impacted by debris accumulation at the culvert inlet.

Calculations based on Manning's equation estimated that the maximum discharge that could be accommodated by the Blackberry Creek culvert was 145 cfs, which fell in the range of the estimated range of the fifty-year RI flood. An analysis of the upstream stormwater culvert indicated that the culvert would surcharge at about 145 cfs, lifting the upstream storm-drain covers and sending escaped flows down Tacoma Street. That explained some of the localized flooding that had been observed at Tacoma and Ensenada Streets, where the overflows were known on rare occasions to collect downstream of the proposed restoration project. A commonly used hydraulic model to estimate flood elevations at the time, HEC-2, was not applied to the project design in this case. Given the limitation of 145 cfs entering the channel with greater flows expected to surcharge through the upstream culvert into the street, adequate channel capacity for large floods in the creek became a nonissue.

At planning meetings for the project, it was pointed out that a tai chi organization regularly performed their spiritual exercises before a redwood tree that perched on the restored creek bank, which meant that project designers had to accommodate not any common, average tree, but a *sacred* tree. Given the intense value attached to the tree, the design was changed to steepened terrace bank slopes for one reach and to decrease sinuosity a little to prevent removal of the tree. The distances between the tops of the terraces on each side of the excavated channel were 45 to 60 feet wide, and the slopes from the bankfull channel to the terrace were excavated at a 2-to-1 slope. The terrace slope next to the large redwood was graded at a 1:1 slope. Using a basic equation for channel length based on channel width from hydraulic geometry data (Leopold, Wolman, and Miller 1964), the channel length should be the range of 70 to 100 feet. The project length of 200 feet allowed for about two 85-foot meander sequences. Hydraulic geometry relations for meandering pool riffle stream types often average riffle spacing at five to seven times the channel widths. Therefore, four riffle rock structures were designed and installed at 40-foot intervals. The straight-line channel slope from culvert invert to culvert invert is 2.6 percent, with the design channel objective to replicate a historic single-thread pool riffle stream type with a slightly lower channel slope attained through adding the sinuosity.

The redwood-tree-constrained right-of-way created concerns for the design team. One was that the channel length would be too short so that headcutting, a response of a channel to flatten an oversteepened slope, could become a concern after the project was constructed. Another was that the redwood tree would be situated on the edge of the right bank with a 1-to-1 vertical slope, which in turn created the need to add a stacked wall of rock from the terrace top by the redwood tree to the stream channel to secure the terrace slope for about 20 feet. The restoration sinuosity is a low 1.1, and the designers believed that it should be closer to the regional averages of 1.2 or more for a slope below 3 percent. To compensate for the lack of channel-length riffle, structures were sized with 1-foot-diameter rock and were buried to act as grade controls and create localized drops to dissipate energy. Rock was also buried in one outside bend about 3 feet back from the boundary of the active channel so that if the channel migrated, it would migrate into the rock and not affect the trail stability at the top of grade.

An additional design concern expressed by the City of Berkeley was that the new downstream culvert and headwall not reduce the 145 cfs inflow capacity to the culvert. Using a head-loss calculation and a conservative 220 cfs discharge estimate for an unimpeded one-hundred-year flow, it was calculated that an additional depth of 3.5 feet was needed above the top of the culvert to support the head necessary to retain the pressure flow of this discharge through the culvert. This project feature was accommodated by a headwall-berm construction at the downstream end at an elevation to support an 11-foot depth from the top of terrace to the culvert invert at the downstream end. Directly upstream, the depth from terrace to culvert invert elevation is 8 feet, and in the midportion, upstream by the existing buckeye and redwood tree, the depth increases to 15 feet. The greatest depth is at upstream reach above the redwood tree at 20 feet.

Because the site is located in a school yard, the project designers needed to address what were fondly referred to as "kid catchers." The grate was designed at the upstream culvert to prevent children from crawling upstream through the culvert. It was designed with vertical bars and hinged to one side for easy opening for maintenance and a locking mechanism for closure. The downstream culvert is grated with a removable steel gate and lock attached to the headwall. The bottom of the grate is tilted up so that the flows will go under the grate unimpeded under most flow conditions, even if some debris catches on the gate.

The preconstruction photo (see fig. 3.16) indicates the stark, playground area that was replaced. Figure 3.18 is a photo of the project excavation, which was carried out by a small business owned by Bill Steele. His rough grading of the site to excavate to the culvert elevation required approximately a five-day work week, with 3,000 cubic yards of soil excavated at a cost of $11,000. Rubble encountered on the site was buried under the trail area. Trucking of the soil to an off-site location was done by a 20-yard truck that took two weeks and 150 loads to remove the excavated material. Figure 3.19 illustrates culvert removal, and figure 3.20 shows the new headwall construction at the upstream culvert. A simple cofferdam made of plywood and sandbags diverted flows around the creek through a pipe. A rock outfall was built for a stormwater pipe on the left bank, and an extra weir of rock was placed adjacent to it for channel stability. The last grading, done by a conservation corps with shovels, was a summer low-flow thalweg channel. Final grading and securing the site with erosion control fabric and installation of soil bioengineering system required ten days and was done by a ten-person Ameri-Corps conservation crew provided by the East Bay Conservation Corps (EBCC).

The construction budget including demolition, grading, culvert grates, bike path, irrigation, and planting, added up to $134,653. Other park wish-list items included a pedestrian bridge over the creek, lighting, and play equipment, putting the project over budget at $232,553. Because the California DWR program would not fund these park features, they were put off for another phase.

FIGURE 3.18 In 1995, the Blackberry Creek playground site was excavated down to the culvert elevation, while saving the existing trees.

FIGURE 3.19 The Blackberry Creek culvert was excavated and cut.

Figure 3.20 New headwalls were constructed around both the upstream and downstream ends of the Blackberry Creek cut culverts.

Landscaping and Maintenance

The design concept was to rely on the use of equilibrium channel shapes and lengths and the use of soil bioengineering to stabilize a newly graded channel without the use of conventional engineering to stabilize the channel. The City of Berkeley had a transparent case of nerves about applying this concept instead of using conventional engineering, and as a result, a couple of the outside bends had a rock toe with 6- to 12-inch rock. The city in part controlled its fears, accepting that a new restoration tool called geotextile fabric was going to be applied to all the graded slopes of the channel, that "equilibrium" channel shapes were being excavated to avoid excessive channel adjustments, and that a certain amount of channel erosion and deposition would help support this stability.

In 1991, we participated in a seminar sponsored by the United Nations International Trade Center, the Coir Board of India, and Robbin B. Sotir and Associates, who brought engineers, geomorphologists, ecologists, and erosion control specialists together to discuss a new product called coir geotextile fabric. The new erosion control fabric being introduced was made of fibers from coconut husks and woven into thin blankets that could be rolled out along road cuts, drainage ditches, hill slopes, or other areas subject to high shear stresses or erosion rates. The professionals evaluating the product recommended its use in concert with soil bioengineering systems. It is organic, it biodegrades after about ten years or more, and it greatly adds to soil strength. The role it was projected to play in restoration was to provide slope stability and time for the plants in the soil bioengineering system to grow, root, and take over as the principle stabilizing method. The manufacture of coir began as a cottage business started and managed by women in India. The husks are soaked and beaten to produce the fiber that was then spun on manual spinning wheels and processed on handlooms or power looms for the final woven fabric (United Nations International Trade Center 1991). By 1995, coir was being distributed by American landscape supply companies, and we used it to secure the open graded site. Conservation corps crews rolled out these "carpets" of coir, aligning them parallel with the channel so that the number of seams subject to the direction of flow was minimized as shown in figure 3.21. Metal staples 6 inches long were hammered into the fabric every 4 to 5 feet to pin it closely to the ground and ensure that there were no gaps or folds for the flows to catch to unravel the fabric. By the late 1990s, coir had become widely available to the growing restoration community.

The conservation corps lined a trench that was excavated along the toe of the channel with the coir fabric. Fascines (sometimes called willow wattles) were made of dormant cuttings from both willow and dogwood, rolled into cigar-shaped bundles, and tied and joined together end-to-end in the trenches. The fascines were staked down in the trench using live stakes crossed over the fascine to hold it in place as shown in figure 3.21. The fascines were covered over by the coir. which is porous enough for plant sprouts to grow through. One of the channel reaches (downstream, left bank) was accidentally excavated too wide. To correct this error and narrow the channel, brush layering was installed as a way to secure a newly filled bank. Brush layering was also used at the upstream culvert outlet to address the high-velocity flows that the channel would receive from the culvert.

Cuttings were made from a mix of willow species (*Salix*), ninebark (*Physocar-*

FIGURE 3.21 Erosion control fabric was installed and fascines were added to secure the toe of slope.

pus capitatus), and dogwood (*Cornus stolonifera*, now known as *C. sericea*). These plants were collected from a nearby regional park in trucks and covered with tarps for transport to prevent the desiccation of the plant material from the windy drive. To reduce plant shock and stress, we waited to collect plant material and install the soil bioengineering systems until the plants had dropped their leaves and were dormant, planting in the last week in October.

At the restoration site, the conservation corps prepared the plant material for installation, manufacturing stakes, posts, and batches of small whip-sized stems. The stakes shown in figure 3.22 were on the average 1.5 to 2 feet long, depending on the species, and were hammered in gently with heavy mallets so that just 1 or 2 inches of plant material showed above the ground. Four-foot long, 2- to 3-inch-wide willow posts, illustrated in figure 3.23, were hammered in along the creek bank's toe margin in areas where rock was placed or in other vulnerable areas. The thinnest willow whips in figure 3.24 were bundled to create the fascines planted along the stream margins. The density of stake planting was high on the terrace slopes, with stakes planted about 2 to 3 feet apart to densely cover the slopes. Trees, shrubs, and ferns were then added.

The planting crews encountered miscellaneous debris and heavy clay soil. No soil amendments were used. A follow-up planting in April 1996 added more con-

FIGURE 3.22 An AmeriCorps member prepares stakes from harvested willow plants.

FIGURE 3.23 Posts, sometimes called poles, are cut and prepared from willow plants and are larger than stakes. Posts are installed in high shear stress areas. They can range from 3 to 10 feet long and 3 to 5 inches in diameter.

FIGURE 3.24 Fascines are made from willow whips that are bundled, tied, and cut square on the ends. They are planted in trenches on the stream slope parallel to the flow.

tainer stock, including shrubs, trees, and ferns. The final project feature was the installation of orange plastic construction fencing along the project margins to reduce damage from children and public visitors trampling the plant material. A substantial irrigation system was installed using spray heads, which the landscape architect checked frequently. Irrigation was used for three years, and photo monitoring indicated that at two years the vegetative corridor was up to 10 feet high.

The plants may have done as well as they did because the ghosts and spooks from the elementary school who frequented the school yard on October 31 (Halloween) as the project was being installed came often to supervise installation. The plants first emerged from dormancy in February, and by April, 95 percent of the willow fascines were growing vigorously. (We started to notice a few years later that dormancy appears to be consistently arriving later in the fall and wonder if that is an indicator of climate change.) The photo records of Blackberry Creek indicate a quick growth from the stakes planted along the channel margin, which were installed to hold the fascines in place. Figure 3.25 shows Blackberry Creek construction after the first rain. Within one year, the channel and side slopes had a complete cover of vegetation, and the channel was well shaded as shown in figure 3.26. The 1997 photos, two years later, show that the channel was lush with growth (figs. 3.27 and 3.28).

FIGURE 3.25 Blackberry Creek was tested with a substantial rain shortly after construction. Note the numerous stakes covering the side slopes.

A field reconnaissance in February, four months after installation, indicated that the plants were breaking dormancy and sprouting. February field notes record that about 60 percent of the ninebark cuttings, 30 percent of the dogwood stakes, and 70 percent of the willows material were sprouting. An April survey found the dogwood performance improving with a 50 percent sprouting rate, with the observation that the dogwood cuttings mixed in with or situated under the willows were the best performers. The presence of shade at this point was adding to performance by 10 percent compared to the exposed areas. On the shady area, 85 percent

FIGURE 3.26 By the fall of 1996, one year after daylighting, the Blackberry Creek channel had a complete cover of vegetation.

of the ninebark was thriving, as was 60 percent of the dogwood. In the sunny areas with little or no shade, the percentages dropped for dogwood to 50 percent and 75 percent for ninebark. Fascines constructed of dogwood were doing poorly, while 90 percent of the willow fascines were growing at this point (Askew 1996).

We were fortunate that the Natural Resources Conservation Service sent us a University of California, Berkeley landscape architect intern, Mimi Askew, to apprentice with us and study for a semester. We worked with Askew to set up four cross-sectional surveys and profile the fall of 1996, which has become our

FIGURE 3.27 This view shows the upstream section of Blackberry Creek two years after daylighting in 1997.

FIGURE 3.28 The downstream section of Blackberry Creek in 1997, two years after daylighting, has the willow and dogwood filling in under the existing redwood and buckeye trees.

project's as-built record. Askew recorded some of the project features in a report that included a project plant list with early survival data that was used later by students and us to record plant vegetation survival over time (Askew 1996; Imanishi 2000; Riley 2013). Within the one-year period, Askew recorded between 70 and 100 percent survival for most container species except *Ribes speciosum*, which had a low 32 percent survival rate, and 40 percent of the ferns died off in the first year. She noted that dogwood-willow mix fascines greatly increased the growth of dogwood in the fascines compared to dogwood-only fascines. She states that her field data indicated that healthy fascines and stakes appear to be a significant factor in the success of other plants and stakes trying to establish on the slopes, possibly by creating a more favorable shaded microclimate (Askew 1996).

Project Lessons and Significance

The Blackberry Creek project was one in which the organizations and firms venturing into the new field of restoration began to get their bearings. They were able to integrate dynamic, functioning streams into a constrained urban environment; advance the application of soil bioengineering into restoration methods; and develop practices on how to integrate local job creation and training and environmental education into projects.

COMMUNITY BENEFITS

The Blackberry Creek daylighting project sent the message that these kinds of projects were safe, even in school grounds, even if the creek was as much as 20 feet below existing ground elevation, and even if the side slopes were steep 2-to-1 and 1-to-1 slopes. Entrenched creek systems with steep slopes create the liability nightmares of city officials, school board officials, and their engineering staffs. This project, though, was well loved, used often, and played a central role in the science curriculum at the school. There are no reports of injuries or accidents at the creek, which is certainly a safer play environment than the playground it replaced. The project served a school demographics composed of 80 percent minority bilingual students including African, Hispanic, and Asian American students. Ray Adams, a teacher at Thousand Oaks School, immersed himself into the project, participating in construction and planting, interacting with conservation corps crews, and bringing his students to the site to watch the project unfold.

Between 2002 and 2004, it became evident that the creek was being polluted with sewage, and signs were set out warning the public of contamination. If the creek had not been daylighted, no one would have known or cared as the pollutants went on their way to the bay. Amy Adair, whose toddler was playing at the creek, contacted the *San Francisco Chronicle*, which featured this problem in the newspaper's "Chronicle Watch, Working for a Better Bay Area" (Vigil 2003). This feature did a nice job of putting local officials on notice of a problem that needed solving and using public exposure and old-fashioned embarrassment to motivate resolution of the problems. By 2004, the *Chronicle* carried a feature in which Berkeley's Mayor Tom Bates expressed admiration for the effectiveness of the elementary school student lobby, which relentlessly pursued local officials to act to clean up the sewage, complete with letter-writing campaigns. Under the direction of the school's science teacher, Jon Bindloss, the daylighting project was turned into a science, aquatic biology, chemistry, and political science hands-on education at the school. In 2004, the daylighting project was back in full public use.

Several visits to the site in 2012 and 2013 indicated that children and adults alike crawl among the jungle of willows, alders, and dogwood in this wild, unkempt tangle of trees and shrubs. Leaning branches have become the banisters that children use to navigate up and down the steep slopes. Students have carefully placed a few rocks in the channel bottom to provide a crossing where more ambitious plans once called for a bridge. Some university student projects to evaluate the creek project were done for class projects in 2000 and 2005, and after interviewing neighbors and teachers, the issue of how deep the creek is and the steep slopes is not mentioned in their reports (Imanishi 2000; Gerson, Wardani, and Niazi 2005).

One of the project's social and economic benefits was to integrate the training and employment of conservation corps youth into the construction. The final grading and planting of soil bioengineering systems is a labor-intensive activity. The advent of the California Conservation Corps in the late 1970s and the growth of this movement to the development of local corps by the 1980s enabled our urban stream restoration movement to start implementing these labor-intensive projects. Our nonprofit organization, the Southwest Coalition to Restore Urban Waters, a regional branch of the national coalition, developed a close working relationship with the East Bay Conservation Corps (EBCC). Because we had ongoing stream restoration work by the 1990s, we put the EBBC on an annual retainer contract, and the corps supervisors gained restoration design and construction experience through participation in the projects. The corps we used on this project was an AmeriCorps crew sponsored by the EBCC. One of the supervisors of this crew, Drew Goetting, was later hired by our nonprofit organization and eventually became its restoration director. By end of the 1990s, both the Urban Creeks Council and the Southwest Coalition to Restore Urban Waters were well populated with staff who had their origins in this conservation corps. Josh Bradt and Steve Connolly from the EBCC supervisor ranks eventually became executive directors of the Urban Creeks Council, and Mike Vukman, another EBCC supervisor, directed Urban Creeks Council restoration projects and programs for many years. The corps became a way that we could guarantee community benefits through employment of entry-level job seekers for those who had graduated from high school or held equivalent degrees. Equally important, these projects served as training and hiring ground for corps supervisors, thereby producing highly skilled, college-educated managers and restoration professionals who have been well positioned to advance in careers in watershed management and restoration.

LANDSCAPE DESIGN AND PLANTING

Because the neighborhood association said that the most important item to them for the creek project was that they wanted the creek to gurgle, Gary Mason, the project landscape architect, implored me to find a way to have the creek splash and make noise "like a brook." The landscape design aspects of this project made us realize that we needed to provide education and awareness of the ecological role of creeks and riparian vegetation to the public situated near these projects. There was likewise no awareness about soil bioengineering and what to expect as it grew. Gary worried that the bushy, dense look of the fascines and cuttings could alienate the neighbors, and he produced a drawing to illustrate the evolution of soil bioengineering plant growth from bushy ground cover to higher canopy trees and a three-layered plant community. Part of our job was to help people understand how soil bioengineering made it possible to uncover a creek from an underground culvert without using the conventional engineering approach of locking the creek in with large rock and concrete. We needed to help them visualize the creek vegetation evolving into different forms over time. The City of Berkeley as well as the Urban Creeks Council helped by interacting with the public on these issues. Eventually, with the building of the tot lot, which was the priority for many neighborhood people, grumbling by the few about a "wild" creek ceased (Schemmerling 2013).

Students from a landscape architecture department wrote evaluations of the project and noted that one influential neighbor thought that a detraction of the project was that she could not see the water in the creek from her house across the street. In their reports, the students took the position that the public concept of attractiveness was an overriding objective to guide stream restoration design. These student reports reinforced that not only the general public, but also the next generation of landscape design professionals, need to be exposed to the objectives of designing for functioning habit, water quality, and in-stream habitat. The designer needs to go beyond being a good listener, merely absorbing the uninformed illusions or prejudices of a public, to determine the form of a restoration project. In addition, the designer must be willing to add the role of teacher to his or her professional responsibilities and describe functional restoration objectives as discussed in chapter 2. Once the public learns that functional restoration design helps birds, fish, and animals and is an urban amenity for humans as well, conflicts that may have surfaced about project features can be addressed early in the process. This statement should not be interpreted as an invitation to not respect and acknowledge public concerns or needs, Rather, it is an attempt to add the element of a better informed public before settling on design elements and to include the discussion that the project does not need to be completely human centric.

Wolfe Mason Associates, the Berkeley Unified School district, and the Urban Creeks Council won an American Society of Landscape Architects Award of Excellence, Restoration and Preservation in 1997. As with Strawberry Creek, the Blackberry Creek project added to the visibility and advancement of the concept of daylighting. I was host to a visiting delegation of Environmental Protection Agency staffs from Washington, D.C., and a public relations astute Cooper's hawk (a riparian species) showed up to be photographed in the buckeye on the creek bank.

Monitoring of plant species survival over time revealed that the corridor supported ten of the fourteen species planted by 2005. The dominant species that flourished at the site by 2013 were the native California blackberry, grown by the California Native Plant Society nursery; the ninebark, arroyo willow, and dogwood collected from the nearby regional park; alders and a few maples; and snowberry and current, which could be found under the shade of the existing redwood and buckeyes. Wild rose was a survivor in sun and shade. Densiometer measurements in 2005 indicated that the canopy cover ranged from 81 to 100 percent (Gerson, Wardani, and Niazi 2005). Photos indicated that the site may have reached this coverage as soon as 1998, or three years after construction. Nonnative species, including English ivy (*Hedera helix*) and a grevillia species, invaded the site. Some native oaks (*Quercus agrifolia*) appeared as volunteers at the site but were located in shady locations, which has prevented some them from thriving. Squirrels could be blamed for this lack of horticultural knowledge. Douglas fir also appeared by 2000 (Imanishi 2000), as did some more redwood trees. It is probable that both species were live Christmas trees planted by the public. A spring follow-up planting in 1996 added wild ginger, two additional species of current, evergreen current (*Ribes viburnifolium*) and golden current (*R. aureum*); redbud (*Cercis occidentalis*); and some chain ferns (*Woodwardia fimbriata*).

The plant species survival list eighteen years later, even after several supplemental plantings were done over time, narrowed to four shrubs species out of twelve planted, six tree species out of ten planted, and no surviving vines or ferns. By 2013, the site had no ferns, redbud, columbine, evergreen, golden and gooseberry currents, or wild grape. The poor performance of the golden and evergreen currents could be because they are not indigenous to this fog-belt coastal environment and are found in drier inland environments. The surviving shrubs were mostly limited to those areas where they were planted under existing canopies of the large redwood tree and buckeye trees. Although the species composition is now lower than was expected, the riparian corridor is lush (fig. 3.29), despite the heavy use of the creek corridor by wandering students. It is apparent that the lack of soil amendment did not hold back the rigorous growth of the riparian corridor.

Minimal maintenance or management at the creek site has been handled by the City of Berkeley, the agency in charge. The city mows the park lawn, attends to trash pickup, and essentially leaves the creek alone. Mason believed that the aggressive early growth of the riparian vegetation had to do with working with the city to monitor the performance of the irrigation system and our planting of dormant plant material for the soil bioengineering, which aided its high survival rates. City employees and the project designers noted the important role of the temporary orange plastic fencing to keep young plants from being trampled. A few people were observed lifting the fence up to let their dogs run around in the protected zone, but the fence did send the message that people were supposed to respect the restoration area, and this strategy was effective for the most part. A project completion celebration was held in May 1996, and we kept the fencing up to remind the attending public that we needed their cooperation to let the plants grow in this sensitive environment. The city kept the fence in place for two years.

FIGURE 3.29 This view shows the Blackberry Creek riparian corridor in 2013.

Carole Schemmerling of the Berkeley Parks Commission supervised some pruning of the riparian vegetation overhanging the trail on the north side of the riparian corridor by the BYA crews. Several visits to the site mostly involved trash pickup. The Urban Creeks Council collected a small amount of willow from the site in the late 1990s, which was most likely the only pruning of plant material at the site. Except for a few volunteer days organized by the Urban Creeks Council members to remove ivy, no organized effort has been put in place to address invasive species control. The ninebark and native blackberry have withstood the aggressive onslaught of the English ivy and still remain the dominant ground cover. The dogwood and willows have also kept exotic invasions at a low level.

The coir fabric appeared to function as a weed-suppression element, but probably the most important control was soil bioengineering, which, as a result of the high density plantings, produced a fast, aggressive cover on the entire square footage of the site within weeks. The coir fabric was a popular feature with the city and public because it gave the site a finished rather than a raw construction site appearance.

STREAM GEOMORPHOLOGY AND HYDROLOGY

The project design probably did not completely meet the public objective of acquiring a babbling brook, but the stream did break up its slope with numerous steps and pools, ultimately creating the public's aesthetic fantasy. The design dimensions for the bankfull channel were close to the postconstruction monitored channel adjustments, but these observations indicated that the design channel was a little too wide and too deep. Askew recorded an immediate adjustment within the first year of a channel narrowing to 6.5 feet from the constructed 8-foot wide channel. She measured adjusted depths from 1.5 feet deep to 0.43 foot. She assumed that the channel narrowing was due to the influence of the dense growth of soil bioengineering on the bank, which was a reasonable hypothesis. A 1999 survey, however, indicated that the narrowing was a long-term adjustment, even with the maturation of the soil bioengineering systems in which banks of dense shrubs were now tree stems with much less dense stream bank plant cover. Surveys completed in 1999, 2000, and 2005 shown in figure 3.30 also indicated a bankfull channel of about 6 feet (Riley and Goetting 1999; Imanishi 2000; Gerson, Wardani, and Niazi 2005). The riffle structures built in a two-step configuration, composed of eight to nine rocks apiece, did not provide the function of stabilizing the profile or, for that matter, the babbling sound many wanted. Some of these structures were partially buried through aggradation near the culverts. A 1999 cross-sectional by the WRI closely matched a 2005 student survey, so most of the channel adjustments occurred within the first four years. The original survey has an error of about 10 feet in the distance parameter, but after correcting for this error, the elevations closely match subsequent surveys. Using cross-sectional data, it appears that the creek aggraded about 1 foot.

Channel profiles surveyed by University of California, Berkeley students in 2000 and 2005 also show a progressive aggradation a few inches to 1 foot from the original design grade and flattening of slope at the upstream and downstream culverts. The upstream culvert grate created a reduction in flow velocities and subsequent dropping of sediment into the culvert above the grate. Sediment accumulation in the upstream culvert and downstream below its grate can explain the upstream aggradation, and downstream aggradation can be explained by the downstream culvert backwater. In between these flattened slopes, the stream slope readjusted with localized steeper slopes, which the student surveys picked up as an increase in riffles and pools from two riffles and four pools in 1996 to nine riffles and seven pools by 2005 as shown in figure 3.31 (Askew 1996; Imanishi 2000; Gerson, Wardani, and Niazi 2005). The stream broke up its slope into locally steepened reaches and evolved into a pool riffle, step pool hybrid channel form.

The lesson we learned was that the final stream form was ultimately going to be dictated by the stream itself. The constructed riffle steps probably provided a

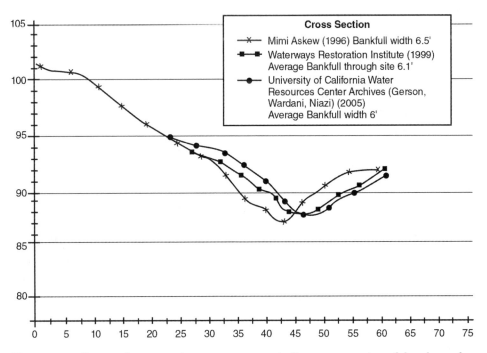

FIGURE 3.30 Surveyed cross sections over ten years indicate a narrowing of the channel to stable dimensions. *Credit: Lisa Kreishok.*

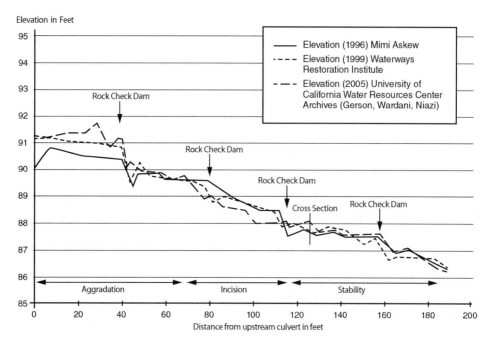

FIGURE 3.31 The Blackberry Creek profile evolved over time, flattening at both ends near the culverts as shown by surveys in 1996, 2000, and 2005. *Credit: Lisa Kreishok.*

stable context for the stream to re-create a thalweg and break the slope up into more elevation drops. After this project, we decided not to construct low-flow channels within bankfull channels, assuming that the stream would quickly likely form its own within the bankfull channel.

A lesson for the urban streams restoration designer is that the sediment regimes affecting the project areas can be unpredictable. By 2013, the upstream portion was so aggraded that the channel was becoming wide and braided in this reach. The safety grate on the culvert has now been left open to allow for sediment trapped behind it to start to transport downstream. The safety grates and sediment transport are in direct conflict with each other. In 2013, I consulted with the Berkeley Public Works Department to see if we could determine the cause of the sedimentation issue. City staff could not identify a specific cause, but their hypothesis is that breaks in water supply lines are common in the area and can introduce sediment into the waterways where the lines cross the creeks, or the discharges can enter the creeks through street flow and storm drains (Akagi 2013). The rock riprap on outside bends was not needed because the stream deposition essentially buried these features and the stable channel slope, good vegetative cover, and equilibrium dimensions have contributed to the stability of the stream channel and its ability to adjust to new conditions.

The discharges over the period of eighteen years have included the large floods of 1996–1997 and 2005–2006. The latter flood was measured by the City of Berkeley hydrologist with a rain gage in her nearby backyard as a 1-in-25-year RI flood, which was typical for other locations for this event in western Alameda and Contra Costa Counties. Student projects recorded Alameda County Public Works Department rainfall-intensity RI data (Alameda County Public Works Department 1995–2005) and estimate that the project site had been subject to approximately a ten-year-interval flood immediately after construction in December 1995. The county record indicates that seven two-year-interval floods occurred between 1995 and 2004, a five-year-interval flood occurred in 2000, and a ten-year-interval flood occurred in 2002 (Gerson, Wardani, and Niazi 2005), and it was subjected to about a 1-in-25-year flood in 2004–2005. The channel has been subject to numerous channel-forming flows and has weathered a significant flood event immediately after construction, before the soil bioengineering systems had rooted, emphasizing the utility of substituting equilibrium design combined with soil bioengineering systems instead of putting urban streams in straitjackets.

Active versus Passive Restoration

Upon making a presentation about this project to a university class, the professor remarked that we should have just opened the culverts and let the stream "restore" its dimensions on its own. The particularly formative experience from the project on nearby Codornices Creek affected the design process we used for the Blackberry Creek project regarding this issue of using passive or active channel restoration methods in urban settings. The Codornices Creek culvert at Eighth Street in Berkeley was excavated and removed using a "rough grade excavation" carried out with the hope that a stable channel would form within the new excavated floodplain. This channel was dramatically unstable for the first two years before we intervened with a follow-up grading project to create equilibrium dimensions. During the two years of uncontrolled adjustments by the creek, a significant area of banks collapsed, and plantings and erosion control measures were washed out. A gas line was exposed as the creek continued to adjust, leading to a rainy-day emergency project with the local utility. The hopes for soil bioengineering systems to stabilize an incorrectly sized channel were dashed. The Codornices Creek channel would have probably remained an erosional, bank-sloughing environment for several more years had we not intervened.

Although it may be reasonable to apply passive, self-adjusting approaches located in a more rural environment in stream corridors not located a few feet from major structures with low hazard areas for public contact and safety issues, the urban environments we were operating in could not tolerate this level of risk. Our Codornices Creek experience, put a nail in the coffin for any future urban stream restoration and daylighting projects using a passive, self-adjusting channel approach because we would lose credibility with the local governments that were responsible for the streams and public safety, the natural resources agencies, and granting programs as well as loose public confidence. Blackberry Creek provided the contrast and experience we needed: a project that took risks in its location in a school yard and design using steep terrace slopes and entrenchment but that was well designed and informed by hydraulic geometry, hydrology, hydraulics, and soil bioengineering. The project context was that it was responsibly designed with the objective to get as close as possible to equilibrium conditions and leave a channel uncontrolled enough so that the stream could adjust and provide the ultimate final stability through slope and dimension changes.

The other formative experience was learning to put together an adequately prepared construction budget. The expenses of hauling off the excavated material from the site and transporting it to a landfill had not been worked out in detail at the time of the grant application. The final excavation and off-haul bill was $11,000 over budget. The Urban Creeks Council ultimately took the financial hit for this cost overrun. The overrun loomed as a constant shadow over the project. Future excavation projects taught us that sometimes toxic soils were involved with requirements to off-haul to expensive designated landfills for hazardous substances. Other surprises involving excavation included removal and hauling of large pieces of uncovered debris. Whereas other construction costs are fairly predictable, we now knew that we would have to practice studied caution with this project planning aspect.

Finally, we learned that bringing a creek back from underground could be celebrated by a neighborhood, school, and commercial area. Figure 3.32 shows how the creek fits into the school and neighborhood environment.

FIGURE 3.32 A bird's-eye view of the Blackberry Creek project in 2014 shows a park with both an active playground and urban creek "playground." *Photo credit: Cris Benton.*

Baxter Creek, Daylighting in a Median Strip, El Cerrito, 1996

Location: Between Poinsett and Rosalind Avenues, downhill of Edwards Avenue, El Cerrito, California
Drainage area to project site: 0.2 square mile
Project length: Valley length: 220 feet 250-foot channel length

Project History

Norman La Force, the mayor of El Cerrito, received a call from a very upset resident in the Poinsett-Rosalind neighborhood. Work had just been completed in 1995 to excavate an old culvert in the median strip between Rosalind and Poinsett Avenues. The construction crews removed the existing culvert and left behind a V-shaped ditch filled with rock. Piles of rock covered portions of the valley slopes along the channel, placed in five butterfly-shaped 20- by 20-foot wedges spaced about 25 feet apart. This startling landscape was not what neighborhood residents expected when they had advocated transforming an old storm drain into a neighborhood creek.

The origins of the Baxter Creek daylighting project (fig. 3.33) began in 1994 when El Cerrito conducted a storm drain and creek restoration study to address structural and capacity issues of its stormwater system (City of El Cerrito 1994a, 1994b). Reports identified a buried 24-inch concrete and brick pipe known to have inadequate capacity as the cause of flooding between Edward and Carquinez Avenues.

Two alternative projects were identified to address this situation. One alternative was to put a new culvert under Poinsett Avenue parallel to the existing culvert that was located under the median strip, and the other was to remove the existing culvert and replace it with an open channel through the median strip. The city estimated the costs of simply replacing the existing 22-inch culvert under the park with a new 33-inch culvert. The median strip is 660 feet long. The upper 375 feet was in turf as shown in figure 3.34, and the lower portion is in a recreational area with restrooms and a basketball court. The area that was most feasible to run an open channel was the upper 375-foot turf area. In the meantime, shortly after these reports were published, the advent of popular creek restoration activities in the East Bay precipitated a Joint Watershed Goals Statement signed by the cities of Richmond, El Cerrito, Albany, and Berkeley and the East Bay Regional Park District, which stated that these entities would establish partnerships to seek opportunities to remove culverts, restore creek habitats, reduce polluted runoff, create open spaces, and promote public awareness of creeks (Joint Watershed Goals Statement 1995).

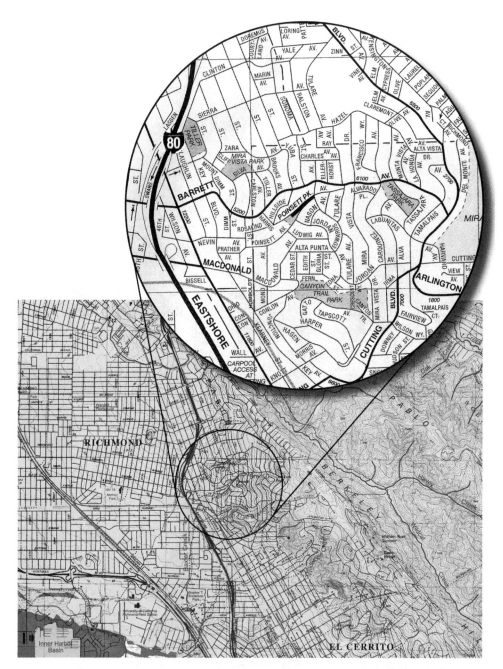

FIGURE 3.33 Baxter Creek drains the cities of El Cerrito and Richmond to San Francisco Bay. The project location is in the midwatershed in a residential setting immediately upstream of a major commercial district on San Pablo Avenue. *Credit: Lisa Kreishok.*

FIGURE 3.34 Baxter Creek flowed underground in a culvert under this median strip lawn between two streets.

There were several challenges to the open channel alternative. One was the steep 10 percent valley slope on which the creek would be created. There was also a topographic dip across the width of the site with a cross-slope differential of 3 to 4 feet been the two streets. The loss of the grassed park and potential loss of views across the park were also listed as possible constraints to public acceptance (City of El Cerrito 1994b; Owens Viani 1996).

The mayor was attracted to the alternative of creating a new creek channel through the grassed park. The city maintenance and engineering manager, Mori Struve, supported this idea because the cost evaluations indicated that the open channel, or daylighting alternative, was significantly cheaper than putting in a new culvert. The manager also found the greater capacity of the open channel an advantage to reducing neighborhood flood damages. Likewise, his experience as a city engineer taught him that his chronic maintenance problems were not open channels but culverts, which could be very difficult to clean of debris. These debris-clogged culverts were the main flooding problem for the city (Riley 1998). The Friends of Baxter Creek, which had formed in 1993 in this neighborhood, had created a following to "green" a nearby stretch of the creek along the Ohlone Greenway and to make a streamside trail along the Bay Area Rapid Transit rail rights-of-way (Owens Viani 2013).

Lisa Swehla and Thais Mazur acquired the city notice for the public meeting for providing public input on whether to daylight or reculvert the creek. Encouraged by the mayor, they distributed the notice door-to-door throughout the neighborhood to encourage their neighbors to attend the public hearing and support the daylighting alternative. A number of neighborhood meetings and a public hearing indicated that although some neighbors were opposed to losing the lawn median strip, there was significant support for the daylighting alternative. The city learned that people who had been in the neighborhood longer than others tended to be attached to the median strip they had always known but were nonetheless convinced that the open channel was cheaper than the parallel replacement culvert alternative. Those with young children tended to see an opportunity for their children to experience and play in the creek (La Force 2013). On July 25, 1994, the city passed a resolution to proceed with the creek daylighting and authorized a grant application to the California DWR to supplement the city funding for the project. Shortly after, a well-organized opposition to the daylighting project formed through the El Cerrito Citizen's Alliance. The alliance tried to head off a grant award to the city, sending letters opposing the project to the DWR and the city council (El Cerrito Citizen's Alliance 1994). The pro daylighting organization prevailed despite this opposition, and DWR awarded a restoration grant in August 1995.

The challenge was to implement a new kind of flood damage reduction project that involved replacing a culvert with a channel. The engineering firm that had completed the stormwater system inventory was tasked with designing this new open channel. This original design used some basic engineering that involved building detention basins at each end of the culvert opening. The upstream basin was intended to slow discharges exiting the culvert, and the downstream basin was designed to provide the depth or head needed to return pressure flow to the downstream culvert. The engineers designed a V-shaped ditch to convey the flows and used scattered riprap for stabilizing the channel. The channel capacity was designed for the 1-in-10-year flood. After this project was installed, neighbors began to question the value of the daylighting, and disenchantment progressed through the neighborhood with the opening of the creek into an industrial drainage ditch (Owens Viani 1996).

The public revolt included the two women who had walked the neighborhood to develop support for the daylighting alternative, who appealed to the mayor and engineering staff at the city to do something to correct the unpopular outcome. Mori Struve, the city's Maintenance and Engineering Services manager who had previously worked with the Urban Creeks Council in 1994 to apply for a grant from the California DWR to supplement his budget for the daylighting project, called the council for help. He was joined by the mayor, who called the local East

Bay nonprofit organizations involved with the successful daylighting projects and urgently requested help. The mayor was worried that this experiment in stream restoration could now set back the concept of daylighting based on the loss of credibility associated with this project. A March 12, 1995, meeting gathered neighborhood and city officials to see if the city could redeem the daylighting concept, and the mayor and city engineer agreed to follow the Urban Creeks Council and Friends of Baxter Creek recommendation to hire a local nonprofit organization with experience in creek daylighting (La Force 2013).

Project Design and Construction

The first project installed was based on the same engineering plans and details used to design stream projects for two dissimilar sites: one was at a much flatter City of El Cerrito Ohlone Greenway location at a 0.015 slope and larger drainage area, and one was for the 0.095 channel slope at Poinsett Avenue. The designs were intended to be open, engineered stormwater channels with 2-foot wide channels, 2-to-1 side slopes, trash racks, and use of stacked segmented concrete wall units to stabilize the side slopes steeper than 2-to-1. The segmented wall units were composed of blocks about 10 inches in diameter and were fit and stacked 2 feet into steep excavated slopes. The walls ranged from 5 to 8 feet tall at 1-to-1 slopes.

At the Poinsett Avenue construction site, a downstream detention basin was excavated 9 feet below grade and 18 feet in diameter. The upstream detention basin was excavated about 9 feet below grade at a diameter of 14 feet. Depths to the channel invert from grade were typically 5 five feet through most of the project. The channel slope at Poinsett Avenue dropped 19 feet over a distance of 200 feet, for a very steep channel slope of 0.095 percent. The site conditions that the WRI found in October 1996 after we were contracted to redesign and rebuild the project contained a channel already eroding around the rock placed in the undersized 2-foot-wide V-shaped channel from dry season low flows.

It was a project that we had to install quickly to address the political pressures to salvage support for the project concept and to complete construction before winter flood flows started. Given the remaining budget for the project, it was paramount that it be accomplished very inexpensively. The WRI worked with the original engineering contractor to produce a new grading plan and agreed on specifications for the step structures. It was also agreed that the WRI would remove the V-shaped rock-lined channel as shown in figure 3.35 and supervise a design-build restoration project that would replace the unstable rock-lined ditch with an equilibrium channel. For regrading the site, the WRI employed Mike Riddle with his family-owned excavator business, who brought in another small family business to haul rock and excavated soil. Large cost savings were realized

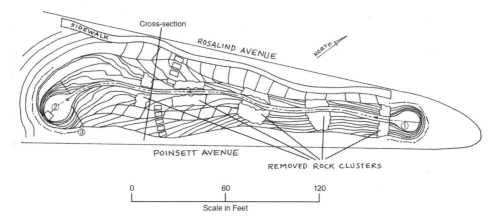

ELEVATIONS:
① 195.5' Bottom of Upstream Culvert
② 174.5' Bottom of Downstream Culvert
③ 182'   Top of Bank

Plan adapted from Waterways Restoration
Institute and Harris Associates

FIGURE 3.35 Rock was removed from Baxter Creek's V-shaped channel, and a wider channel and floodplain were graded. *Credit: Lisa Kreishok.*

by identifying a community garden in Richmond that wanted the excavated soil. In addition, another city public works project had use for the rock removed from the site. The total design and construction budget that the WRI had to work with was $7,500 from the state grant and $1,000 from the engineering contractor.

Reference information to inform the project design was spotty, with one or two backyard locations providing us with the perception that an active channel would probably adjust in the range of 5 to 7 feet wide. The only reliable information we had on channel dimensions came from the regional hydraulic geometry curves, which indicated a cross-sectional area of about 8 square feet with a width of up to 9 feet. Regional regressions were used to estimate a 1-in-10-year discharge of about 85 cfs, which was the city's target flow for channel capacity. We selected channel design dimensions at 6 to 7 feet wide and 1 foot deep based on our observations that the channels we were constructing in the East Bay were filling in at a lower cross-sectional area than the average values indicated by the Bay Area regional curves. The watershed drainage area was similar to the Blackberry Creek restoration just completed for a 0.3 square mile watershed. The Blackberry Creek channel had adjusted to lower dimensions than the regional curve and had not experienced excessive erosion from high water in 1995, and this experience also helped inform this decision.

The final design features were to create a 250-foot-long creek channel within a project right-of-way width ranging from 35 to 84 feet, grade back the stream banks'

3-to-1 slopes, and create a floodplain to provide a safer, more stable environment for the neighborhood. The design in figure 3.36 illustrates the grading plan and removal of almost all the rock used in the first project. A small portion of the rock was kept on site and used sparingly to build the step pools. The bank stabilization emulated the Blackberry Creek project by using a fascine along each toe of slope composed of both willow and dogwood cuttings, coir fabric, and staking of willows and dogwood along the side slopes.

Although we knew that the channel should be composed of steps and pools for this steep valley gradient, there was no guidance on the design of step pool channels available at this time. The WRI spaced six step structures located at seven times the bankfull width or about 30 feet apart using the only known guidance that is applied to pool-riffle stream types associated with flatter gradients than this step pool stream. The design detail for construction of the steps in figure 3.37 borrowed from the check dam concept, with one-quarter-ton rock boulders buried

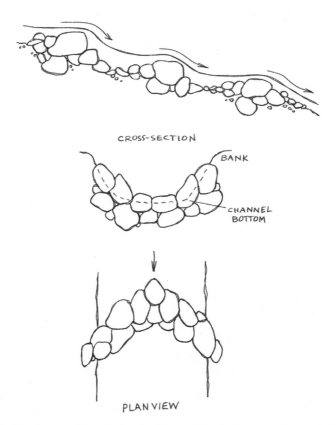

FIGURE 3.36 The final restoration design for Baxter Creek designated a graded floodplain, 3-to-1 channel side slopes, and a step pool channel to break up a 10 percent slope. *Credit: Lisa Kreishok.*

two-thirds below grade; the top of the rock helped form the desired design slope, which evened the fall from the upstream and downstream basins. Smaller boulders were fit up against the toe of the larger rocks, which were all placed perpendicularly across the channel. The downstream rock forming the step was placed at a lower elevation to produce a drop ranging from 6 inches to 1 foot. The spacing was based on observations of constructed step pool channels in the Strawberry Creek watershed on the university campus, some of which became unstable if more than 1 foot in height from the force of the overstep flows.

Construction was accomplished in nine days between November 5 and 15, 1996. The WRI had the fortunate situation of employing the East Bay Conservation Corps (EBCC) under a retainer contract and had some credit in its account. As fortunate fate arranged, a variety of rock sizes were in the load to be removed from the site, and WRI had a wide range of available rock sizes to build the steps. Smaller rock was fit into the large rock structure as it was constructed. Drew Goetting, now with the WRI, supervised the corps members to build the steps and pools. Figure 3.38 shows the spacing of the rock piles for constructing the steps. Six rolls of coir fabric were stapled into the trench following the toe of the channel where fascines were placed. Eight rolls of less expensive straw fabric were used to stabilize the upper slopes. A rough grading of the site was accomplished with

FIGURE 3.37 The concept design guiding the design-build step pools employed a range of sizes of boulders and cobbles. Most of the rock in this photo was removed.

the excavator, and the conservation corps did the final grading with shovels and installed the fabric and soil bioengineering (fig. 3.39). In some areas, the channel was excavated slightly too wide, at 8 to 10 feet. The fine grading created the continuity in channel shape, and a slightly overwiden area near the downstream end of grading was rebuilt using brush-layering soil bioengineering systems. Willow posts were integrated into the edges of the step pool structures. Combining the fine grading and soil bioengineering systems effectively narrowed the channel to a 6-foot width. The project used eight three-quarter-ton pickup truckloads of collected plant material from the nearby regional park.

The first substantial rain of the season occurred between November 15 and November 17, immediately after project completion. Monitoring the site indicated that we needed to add more steps for stability. We used the material on site to build two more steps and rebuild one existing step, for a total of eight steps completed toward the end of November. The first rain was already redistributing some of the material in the steps, breaking the slopes into smaller incremental drops. The diversity of sizes of rock incorporated into the steps had provided for the wide distribution of material over the entire channel profile. Container alders were added to the margins of the channel to add stability to the steps at this time. Container bigleaf maples were added into the willows to provide for a later-

Figure 3.38 Some rock was salvaged from the first engineering project on the channel to build the step pool channel at Baxter Creek.

FIGURE 3.39 The East Bay Conservation Corps completed the final grading, fabric installation, and soil bioengineering for the Baxter Creek project.

developing canopy species. A January 1, 1997, twenty-five-year recurrence interval flood continued to disperse and rearrange material between the steps. Shown in figure 3.40, this self-organizing channel has provided a stable channel, and as of 2015, no flood overflows have been experienced in the streets since the project was constructed.

The neighborhood was relieved to see a channel with natural dynamics, riparian vegetation, and flood damage reduction for the greater-than-ten-year flood. The $7,500 cost of the reconstruction included project design, on-site supervision,

FIGURE 3.40 Transport and re-sorting of the mobile rock and cobbles by flood in early January 1997 began to rearrange the steps and pools of Baxter Creek.

and $3,800 for excavation and hauling. An additional contact between the city and the conservation corps added $2,500 to the budget. The original excavation and culvert removal was $10,000. A landscape architect was hired for approximately $5,000 in 1997 to landscape the margins of the site, bringing the total project cost to $28,800. A concrete path and boulders arranged on the north end of the project added another $26,000 in cost on top of restoration costs. The original city budget for removal of the 24-inch culvert and reinstallation of 33-inch culvert through the same alignment in the median strip was about $39,000 (Struve 1998).

Landscaping and Maintenance

By the spring of 1997, a substantial corridor of vegetation was established along Baxter Creek from the soil bioengineering as shown in figure 3.41. In 1997, a landscape architect hired by the city planted chaparral species along the margin of the site to add color and an aesthetic flourish ringing the riparian area. They included California buckwheat (*Eriogonum fasciculatum*), purple sage (*Salvia leucophylla*), California fuchsia (*Zauschneria californica*), sage leaf rock rose (*Cistus salvifolius*), ceanothus (*Ceanothus maritimus*), and coyote bush (*Baccharis pilularus*). Some riparian species were added into and next to the riparian corridor already with well-established soil bioengineering, including buckeye (*Aesculus californica*), coast live oak (*Quercus agrifolia*), California hazelnut (*Corylus cornuta*), dogwood (*Cornus stolonifera*), coffeeberry (*Rhamnus californica*), red-flowering currant (*Ribes sanguineum*), California blackberry (*Rubus vittifolius*), clematis (*Clematis ligusticifolia*), California wild rose (*Rosa californica*), snowberry (*Symphoricarpos alba*), and California huckleberry (*Vaccinium ovatum*). The shrubs were planted as 1-gallon plants, and the trees 5 gallon. The neighbors added their own plants along the margins, including nonnative lantana, mallow, and the native sycamore (*Palatanus racemosa*) and ninebark (*Physocarpus capitum*).

FIGURE 3.41 By April 1997, the soil bioengineering provided a stabilizing and aesthetic riparian corridor for Baxter Creek.

Two years later, in 1999, the riparian corridor had occupied most of the median, overcoming many of the landscape architect's plants on the margin and attaining a height of about 15 feet as shown in figure 3.42. By 2012, an inventory of the established plants indicted that most of the chaparral plants had not survived; the exceptions were one surviving rock rose, one baccharis, and one salvia, all on the sunny south side. A few riparian species thrived, including willow, dogwood, ninebark, and native blackberry, all of which were competing with invading nonnative ivy. Oaks and the bigleaf maples were having a difficult time coming through the canopy of the willow and alder. The oaks were trying to twist away from the shade. The buckeyes planted on the edges of the system were thriving.

FIGURE 3.42 By 1999, the Baxter Creek riparian corridor was 15 feet high.

Three basic shrubs succeeded, including coffeeberry and a few currants and wild rose in the north-side shading. The clematis vine was thriving. The plant species survival observed at this location was consistent with other urban creek planting projects in the neighborhood case studies. Chaparral or drought-tolerant species planted for aesthetic effect did not survive well, whereas several riparian species dominated. The predictable invasion of English ivy (*Hedra helix*) and Himalayan blackberry (*Rubus discolor*) had started.

This project was intended to be low maintenance except for the need for weeding and management of plants to allow for later succession canopy development over time (Waterways Restoration Institute 1996a, 1996b). The neighborhood conducted a sustained effort of stewarding the creek and park, adding plants over time and weeding. The Urban Creeks Council, with permission from the city, occasionally pruned vegetation from the corridor to help support species diversity and moderately thinned the willow stands or removed willows structurally failing to use in other restoration projects. This model of maintenance worked for the city and neighborhood but came to a halt in about 2002 when a tree fell on an overhead power line and cut off electricity to a resident. A standoff ensued between the power company and the city over which entity had the responsibility to address the problem.

Eventually, a tree-trimming company was hired, although, according to the city, exactly who did the hiring was unknown. A very insensitive vegetation pruning job was then done, which sent phone calls to the city and the WRI to address the damage. The neighbors most engaged at this time had become attached to the forest and the screening it provided and valued the blocked sight lines across the median, which preserved the feeling of privacy. This group was quite stricken by the chopped-up appearance of the riparian corridor. The city debated hiring a nonprofit organization with insurance and appropriate heavy equipment such as chainsaws and trucks or hiring a full-time maintenance profession to handle all similar park and open space environments. A protest against these options by a citizen's organization that liked to organize volunteer parties to remove weeds and exotics put the city into political conflict, and an impasse ensued. The neighborhood was requested not to do volunteer maintenance until the issues were resolved. The maintenance issue had turned into a ten-year unfortunate, unforeseeable quagmire that did not resolve itself until 2012.

By 2012, the City of El Cerrito had the necessary resources to hire a watershed management expert for their public works department. Stephen Pree ultimately resolved the maintenance impasse by hiring an arborist and evaluating the needs of the trees in the riparian corridor. He also organized an after-work neighborhood meeting in August 2012 and listened to the input of the neighborhood on future management needs. I attended this meeting and was encouraged by the fifteen or so participants. They walked the project area together and discussed the needs of the creek, birds, and frogs. It was interesting to note that fifteen years after the project had been completed, some of the self-identified early opponents to the creek restoration alternative attended this meeting and now expressed full support for the project. Part of the reason for this change of heart came from having fifteen years free of flood damages because in prior years, flooding damages had been chronic. Another reason is that they had come to appreciate how the creek environment enhanced their neighborhood. The consensus direction from the neighborhood was to keep the "wildness" of the riparian corridor and to protect the view screening between the two streets. The participants also requested that the neighborhood be allowed to continue its stewardship projects. A neighborhood landscape architect, Chris Else, drew up plans to add new plants to the park. The city is now sponsoring neighborhood-based work days on the creek.

The subsequent pruning that was overseen by the city was very sensitive to the neighborhood input and helped protect a layering of vegetation with ground cover, shrubs, and canopy trees. Pruning was executed to access more sunlight for the bigleaf maples, oaks, and buckeyes that are now becoming healthier components of the riparian forest. Willows, which had been previously topped and pruned up too heavily, had become top heavy and were structurally weak. The arborist's evaluation noted that generating sprout growth toward the higher part of the trees by removing lower limbs makes the trees more prone to failure. The arborist also noted that selective removal of willows over time to provide space for other canopy species was a better strategy for management than willow pruning. In addition, he noted that one of the willow species used, the yellow tree willow (*Salix lasiandra*), which can achieve heights of 30 feet, may have been out of scale for this small site. He recommended the use of smaller-structured willows for certain urban situations such as *S. scouleriana* and *S. exigua*, which grow to 15 feet (Batchelder 2011). Many restorationists report using sandbar willow (*S. hindsiana*) or red willow (*S. lasiolepis*) when a smaller, more flexible plant is desired.

Project Lessons and Significance

Despite the "rocky" start to this daylighting project, after its redesign and reconstruction in the fall of 1996 it became one of the most visited urban creek restoration projects in the East Bay. The neighborhood wholeheartedly adopted the project and organized planting projects. It became the cover photo for the 1998 Island Press book *Restoring Streams in Cities* (Riley 1998). It is the site for a research project on water quality and social issues around urban streams and for geomorphic research on how step pool channels form. It also became a location where urban stream restoration project maintenance became an issue, with the city achieving a final resolution by hiring its own staff to address city park and creek management needs. Finally, it inspired a host of other restoration projects in the watershed. To its credit, the engineering firm that initially struggled with the first creek opening and was new to the field of restoration turned this project into a learning experience to learn more about the field of restoration.

The Baxter Creek Poinsett Avenue project attracted three different research projects. One involved a professor from Texas Agriculture and Mining University, Anne Chin, who wanted to find out if there was a pattern of step pool development that could be quantified to provide guidance for designers of steep pool channels. Another involved student research by Alison Purcell and others under Professor Vincent Resh at the University of California, Berkeley focused on changes over time of the benthic (bottom-feeding) populations at an urban stream restoration site. Chapter 4 will add the discussion of a third, a bird population study conducted by the San Francisco Bird Observatory that looked at the connection between bird species presence and urban stream restoration projects.

GUIDANCE ON STEP POOL DESIGN

Professor Anne Chin started visiting this site in 2005 to evaluate how the step pool channel morphology developed, which later informed her about how the steps formed themselves in a way that natural stream systems do along the California coast. She added her profile and cross-sectional surveys to existing project surveys from 1996, 1999, and 2002 to evaluate the spacing and height of the steps formed by creek adjustments. Between 1996, when the first eight steps were installed, and 2005, Chin counted the development of twenty well-developed step pool sequences. Figures 3.43 and 3.44 show channel adjustments over time in surveyed cross sections and profiles. This information was combined with data collected from other stream restoration sites and sites in natural, unmodified states along the California coastal region. Chin's research provided mathematical relations on step heights and spacing that are now in use by restoration professionals. She concluded that the Baxter Creek project, which employed a diversity of cobble and boulder sizes and a stream-based distribution of the material over the profile, produced a good model for restorationists to use as opposed to design strategies that relied on more immobile design features such as weirs (Chin, Anderson et al. 2009). Figure 3.45 shows the step pool system performing under the flood flows of January 2006, a 1-in-25-year flood.

In the meantime, I had communicated with practitioners in the Feather River watershed about their experience with designing step pool channels. The Plumas Corporation, a rural-based nonprofit organization, was employing a restoration strategy to rewater meadows in heavily grazed and channelized headwater stream systems referred to as "plug and pond." The projects required using oversteepened

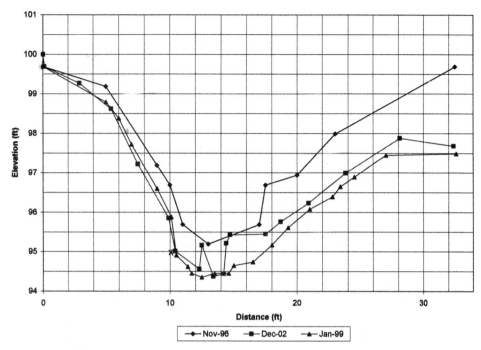

FIGURE 3.43 Cross sections were surveyed on Baxter Creek over time. Graphic credit: Lisa Kreishok.

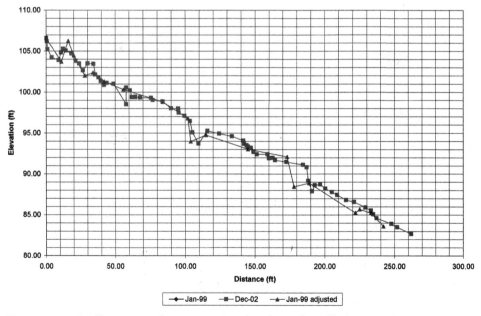

FIGURE 3.44 Profiles surveyed on Baxter Creek indicated a self-adjusting slope over time. *Credit: Lisa Kreishok.*

FIGURE 3.45 The step pool system performed well during the January 2006 Baxter Creek flood.

channel segments to join raised channels with existing lower-channel elevations. The designers were using a similar strategy of stabilizing steep-slope channel segments by introducing a diversity of gravel, cobble, and boulder sizes to the slopes and allowing the flows to entrain and disperse the material to break up the slope "naturally."

The design at Baxter Creek was a happy coincidence in which the material left at the site after removing most of the engineered rock fill from the V-shaped channel provided a good quantity of rock of appropriate sizes, which was perfect for transport and dispersal along the channel. Chin's work is some of the most important applied geomorphic science developed for practitioners because it fills an important gap in which there has not been good science-based guidance on how to assist the formation of stable channels at greater than 3 percent channel slopes. This creek has been very stable located on a 10 percent valley slope, which approaches the level of slope you would expect in a high-elevation-watershed cascading stream type.

The project features that have not been stable over time are the concrete blocks that had been installed to stabilize steep slopes. A local newspaper article that covered the appearance of the "new creek" called attention to the failure of these concrete blocks after the first rains. The city has had to retrieve these and reinstall over time. In hindsight, soil bioengineering systems for the steeper slopes could have provided deeper, better anchored stabilizing systems.

RESEARCH ON BIOLOGICAL AND HABITAT VALUES OF URBAN STREAM RESTORATION

The ultimate compliment on the wildlife habitat values has been the twice-observed visits by a mountain lion to the site. The neighborhood reaction has been one of wonder and appreciation and to leave the lions to their wanderings. The sobering aspect of these observations is the interrupted habitat along the creek corridor, shrinking the area for these normally reclusive creatures. The Baxter Creek site has supported native tree frogs, and chapter 4 goes into detail about a bird population study at the site that followed a bird list developed by one of the neighbors who began recording bird species in 1999 (Owens Viani 2000b).

In 1999, a benthic insect study was started at this site to address the question of whether an urban stream restoration project could increase the biological health and diversity of benthic insect populations (Purcell, Friedrich, and Resh 2002). The benthic sampling compared two sites upstream in unrestored reaches against the restoration site using benthic insect biological indicators to represent differences in water quality. The project also selected the south branch of Strawberry Creek on the University of California, Berkeley campus as a baseline for representing good urban stream conditions. Strawberry Creek had undergone extensive improvements in reducing pollutants in the 1980s. The physical changes to the Strawberry Creek watershed involved building a redwood crib wall about 30 feet in length on a bank erosion location, adding some check dams in several locations, and repairing several gullies in the upper canyon. The operative restoration in this case was the attention paid to the pollution cleanup, not restoration of the geomorphic structure, dynamics, or functions of the creek. The restoration activities addressed the cumulative water quality impacts from up to a hundred pipes entering the creek, some of which involved cross connections between sanitary and stormwater lines, deteriorating sanitary sewers, and discharge of hot cooling waters (Charbonneau and Resh 1992).

The Poinsett Avenue site was compared against Strawberry Creek using a widespread rapid bioassessment method developed by the US Environmental Protection Agency for national use, particularly by volunteers and schools. The benthic organisms are recorded for family richness, numbers of insect families and taxa richness, or numbers of species, and the numbers of the more pollution-sensitive orders are compared with the total numbers of individuals collected. This system provides a quantitative biotic index. The study also uses some benthic counts collected in 1997 after the restoration project construction for comparisons.

A 1999 assessment found that the restoration site was being recolonized and that the macroinvertebrate family and taxa richness were similar to Strawberry Creek. The researchers found better biological conditions at the restored site compared with the unrestored upstream reaches upstream. The study also found that all the unrestored sections of Baxter Creek had more pollution-tolerant species than did Strawberry Creek and the restored Baxter Creek section (Purcell, Friedrich, and Resh 2002).

We were fortunate that Alison Purcell was able to conduct follow-up evaluations of the site in 2004. She found no improvement in the biotic assemblage at the restoration site compared to 1999 and found that taxa and family richness had dropped some from 1999. The 2004 sampling of benthic insects in the restoration reach had higher taxa and family richness than the unrestored reaches, but the Strawberry Creek reference site was found to have higher taxa and family richness as well as fewer pollution-sensitive species. The research concludes that the results from the 2004 biological assessments ultimately indicated no significant improvement in the biotic assemblage compared to 1999. Purcell lists some potential reasons for this finding, including the urban stormwater conditions and the short length of the project as a limiting factor on mitigating pollutants. Purcell ends her study by mentioning that neighbors reported a kill of the large population of native tree frogs and that they believed that potable water with chloramines had been spilling into Baxter Creek (Purcell 2004). The study does not attempt to make links between these reports and the benthic assessment results.

By 2008, researchers had resampled macroinvertebrates at Baxter Creek in part to understand how a step pool morphology affects benthic insect populations. The researchers reported that the unrestored and restored reaches now share the same biological conditions and that the more pollution-sensitive species are found on the step structures compared to the pools (Chin, Purcell, et al. 2009).

Another aspect of the assessment was to estimate the habitat quality of the two different sites, Baxter Creek and Strawberry Creek. The researchers turned to habitat assessment methods contained in the US Environmental Protection

Agency's Rapid Bioassessment protocols that include both a benthic diversity index and a habitat quality rating. The Strawberry Creek site received higher scores than Baxter Creek using the assessment method for habitat (Chin, Anderson, et al. 2009).

How Do We Measure the Effectiveness of Restoration?

These interesting water quality and habitat assessments on Baxter Creek lead us to question how to interpret them. Both the 2004 and the 2008 research mentioned above are cited frequently, along with some other urban stream benthic assessments studies that offer conclusions that benthic populations cannot recover through physical urban stream restoration projects. Left unanalyzed was why the benthic insect populations went through a recovery in the first years after the restoration, only to decline later. Do we conclude that urban streams inherently cannot support biological health? Although the Baxter Creek environment has a substantial multilayered riparian corridor and the channel is recognized to be a complex dynamic environment capable of restoring a step pool sequence, it nonetheless achieves relatively low habitat quality ratings. Its rating from the rapid habitat assessment is low compared with the Strawberry Creek site located in a university campus. How should we interpret this information? Both issues need to be considered in the broader context of what is affecting a small-scale restoration project on Baxter Creek and how we measure quality habitat.

In trying to develop an understanding of the larger context for evaluating the effectiveness of the Baxter Creek project for biological improvements, the WRI collected information on the watershed conditions that are likely affecting this site. By 1999, the East Bay Municipal Utility District (EBMUD) began to convert from treating its potable water supplies with chlorine disinfectant to using more residual chloramines that do not break down as easily as chlorine. Neighbors reported the extirpation of frogs that occupied the Baxter Creek restoration site with the advent of water flows down the street from breaks in water mains to the regional water pollution agency, the San Francisco Bay Regional Water Quality Control Board. It took until 2005 to understand the scale of regular accidental chloramine discharges to East Bay creeks in general. After fish kills in Codornices, Strawberry, and Sausal Creeks, a series of three public meetings were organized with the EBMUD at the request of the Urban Creeks Council and the San Francisco Bay Regional Water Quality Control Board to develop an accounting of the degree to which unplanned discharges of chloramine-treated water was entering storm drains and creek channels and killing aquatic life. The water board had, in 2001, started to require that the EBMUD submit records of its unplanned discharges of treated drinking water from breaks in water supply lines. The submitted records indicated an average of ninety-nine such breaks a month in the EBMUD service area, and maps locating these breaks indicated a concentration in the upper Baxter Creek watershed influenced by movements of the Hayward fault.

By 2013, the San Francisco Bay Regional Water Quality Control Board was working to put in place a Bay Area–wide chloramines discharge permitting process, recognizing that drinking water, sadly enough, is a pollutant that must be regulated and reported. Any urban area that has experienced the conversion for treatment of drinking water supplies from chlorine to the more powerful and residual chloramines must now factor in this new reality if we are to understand the results of our biological assessments.

Future bioassessment sampling could be performed after the now-ongoing water line replacement improvement projects are completed in the upper Baxter Creek watershed. The context for these future studies could employ research questions such as, given the cessation of chloramine discharges, can Baxter Creek compare with Strawberry Creek? Is it possible to do bioassessment comparisons among creeks if we do not have the record of chloramine discharges to the different creeks? If correlations between recorded chloramine discharges and benthic population declines and recovery do not show strong relationships, are there other pollutants, including pesticides, affecting benthic populations? If we can control for the influence of pollutants, can we then ask if there are limiting factors about the Poinsett Avenue step pool project that cannot support a healthy benthic diversity?

This case clearly highlights the confounding variables we need to consider when designing research and making conclusions about the performance of restoration projects using bioassessments as an evaluation tool. Generalizations about benthic populations may not actually serve as measures of restoration project performance because a number of factors may overwhelm even a well-functioning ecosystem to compensate for pollution. Chapter 4 provides more detail on the issue.

The Baxter Creek at Poinsett Avenue case addresses another topic of interest about urban streams: How should we evaluate the structural habitat value of streams? The assessment based on habitat condition used to compare Baxter Creek with Strawberry Creek is controversial, as introduced briefly in chapter 2's discussion of different points of view on whether to use "condition-based" or "process-based" riparian assessments. The simple rapid assessment, used to compare two sites against each other, may have value for large-scale regional snapshot indicators of the health of a region's streams. A number of river, wetland, and riparian scientists, however, find that these condition-based assessments are not well matched to determine the value or effectiveness of riparian habitat restoration projects. A further developed riparian habitat rapid assessment method in 2007 that replaced the assessment used to evaluate Baxter Creek heightened the controversy over expanding these rapid assessments to riparian environments, which were originally developed to characterize wetlands.

The rapid assessment used for Baxter Creek evolved to the California Rapid Assessment Method, called CRAM, which does not recognize nor record in its numerical assessment score that the Baxter Creek environment underwent improvement as an urban drainage in an underground culvert to a free-flowing dynamic step pool channel. A channel constrained within a median strip in a neighborhood setting cannot compete against a channel located on a vast university campus such as Strawberry Creek for an assessment score representing the open space adjacent to the riparian environment. The Baxter Creek project may have achieved its maximum attainable habitat value in a median strip, but this context is not represented in the rapid assessment either.

The rapid habitat assessments such as applied to the Baxter Creek study should receive an updated review by stream scientists who specialize in riparian and stream functions and processes (Hruby 1999; Kusler 2004; San Francisco Bay Regional Water Quality Control Board 2007). The assessment applied in this case in 2002 as well as CRAM dated 2007 (with ongoing revisions) is strangely biased toward an idealized pool riffle habitat, whereas a step pool habitat type is assigned lower numbers in the rating system. Other factors, such as flow regime (how much flow is available year around), are controlled by given watershed factors and therefore should not be par t of a rating factor that measures the performance of restoration projects (San Francisco Bay Regional Water Quality Control Board 2007). Another unfortunate comparison made in this instance was to compare the benefits of a reach-scale physical habitat restoration project against a larger-scale and extensive pollution cleanup restoration project involving many acres. A reach-scale physical restoration project should not be considered equivalent to a watershed pollution cleanup "restoration." If we compare these two very different types of "restoration" and use bioassessment scores to compare the results, the evaluation is certainly predetermined to indicate that small-scale physical restoration cannot somehow outperform the cleanup of actual pollutant discharges.

What assessments should we use to determine the value or results of urban creek restoration? One strategy is to consider whether project objectives were met. One project objective for this site was to reduce flood damages and risks. The performance of the project in flood flows of 1997 and 2005–2006 exceeded the flood risk reduction objective of protection from the 1-in-10-year flood by avoiding damages for 1-in-20 to 1-in-25-year floods. The project was intended to solve a flood problem inexpensively and expeditiously while garnering neighborhood acceptance. An attitudinal statistical survey conducted as part of the 2002 study found that a vast majority were "pleased with the project," that the neighborhood took ownership of the site, and that the project meets the primary objective identified by the neighbors, which was that it "should accomplish the rejuvenation of the native biology and landscape" while also serving as a neighborhood park (Purcell, Friedrich, and Resh 2002). Figures 3.46 and 3.47 show a neighborhood park that changes the character of the neighborhood; compared with figure 3.34, a significant wild environment has been introduced into a very urban setting. Chapter 4 describes a bird population study for evaluating the wildlife habitat of the site, an assessment tool that has potential as a more functional measurement of habitat quality. Water quality measurements taken at the Poinsett Avenue by the water board in 2013 indicated low temperatures and high dissolved oxygen, easily meeting the regional basin plan objectives for these water quality parameters. At one point, the restoration project functioned as a native tree frog habitat. It is hoped that this function will return in the future.

Related Projects

The Poinsett Avenue Baxter Creek project inspired another stream restoration project that was completed by the Urban Creeks Council in 2000. This project was situated in Booker T. Anderson Park, a residential neighborhood city park named after a community leader and city council member near the edge of San

FIGURE 3.46 The Baxter Creek restoration project changed the character of the neighborhood.

FIGURE 3.47 A bird's-eye view of the Baxter Creek project shows a "wilding" of the neighborhood. *Photo credit: Cris Benton.*

Francisco Bay in a largely African American neighborhood of Richmond. This well-loved park can be impacted by crime, and balancing the needs of the environmental restoration projects with the need for safety was successfully resolved. The Booker T. Anderson restoration site on Baxter Creek in turn created a positive example for the City of El Cerrito. The city eventually replaced plans for a large commercial development along a major commercial route with a project to restore Baxter Creek and to create a park that serves as a new gateway to the city, the Gateway Baxter Creek project. The positive community response and experience with these projects motivated the City of Richmond to continue to feature restoration of Baxter Creek at two new sites, the Ohlone Gap greenway located in the commercial district and at the Mira Flores housing redevelopment project. Mira Flores is a brownfield cleanup of a historical nursery, providing a low-income housing project located near regional transit that meets the state's definition for a greenhouse gas reduction development.

Baxter Creek at Booker T. Anderson Park, Richmond, 2000

Location: S. 47th Street near Carlson Boulevard, Richmond, California
Drainage area to project site: 0.92 square mile
Project length: 720-foot valley length, 1,000-foot channel length
Park acreage: 22 acres

Project History and Description

Lisa Owens Viani and her dog, Isis, were walking thought their neighborhood when they came upon the construction site where Baxter Creek was being daylighted in Poinsett Park. She had worked with her neighbors to support the project and was thrilled that it was under way, creating a dynamic, green environment. Owens Viani, a retired ballerina who had moved from San Francisco to the East Bay after the 1989 Loma Prieta earthquake, became engaged with her new neighborhood and founded the Friends of Baxter Creek with her neighbor friend Maryann Aberg. Baxter Creek would have remained an unnamed watercourse at the time of the Poinsett daylighting project if Owens Viani and Aberg had not named the creek and called public attention to its potential value to the community. The Friends of Baxter Creek scoped out the potential creek restoration opportunities in the Baxter Creek watershed using topographic maps, historic maps, and storm-drain maps. Key to all the Baxter Creek projects being completed between 2000 and 2015 is a historical report of the creek and inventory of sites where the creek could feasibly be restored (Owens Viani 2000a).

This group began developing awareness with public officials and the public about the prime opportunity that the open space along and under the elevated BART tracks presented for creating a "there-there" (to borrow a line from Gertrude Stein) or a defining feature of the city. A section of remnant creek ran along this unappreciated weedy space next to a Lucky's grocery store on the busy thoroughfare of San Pablo Avenue. The Friends of Baxter Creek were fighting to save this site as a green space in an area greatly underserved by parks and open space, an area planners had targeted for big-box stores. The frustration of Owens Viani with trying to achieve something at this site since 1993 was harnessed into a new idea: restore Baxter Creek in the existing downstream Booker T. Anderson Park in Richmond, the adjacent city. Her strategy was to create another example like the Poinsett Avenue project and continue to build political support for the BART site centrally located in El Cerrito.

Owens Viani had met the Urban Creeks Council through the Poinsett Park experience and joined up with them for this project. She organized multiple community meetings at the Booker T. Anderson community center and volunteered at the nearby Stege Elementary School. The San Francisco Foundation supported her community outreach work in this mostly working-class, African American stronghold that was diversifying with Asian and Hispanic families. Tony Norris, parks director for the City of Richmond, became an important ally and participant in the project, and Richmond Mayor Irma Anderson (the park was named for her late husband) and Richmond City Council member Tom Butt provided support. With this community support behind her, Owens Viani acquired grant funding from the California Coastal Conservancy and the California DWR to restore a much-degraded creek in the park. Figure 3.48 reveals the very degraded condition of the creek in the park.

Drew Goetting of the WRI and Josh Bradt of the Urban Creeks Council assisted Owens Viani with stream restoration design, and this project developed into her master's degree project in geography for San Francisco State University. Bradt and Owens Viani used historic data on the creek alignment (the creek was ponded for a frog-raising business in the early 1900s), regional curve data, and region-based hydraulic geometry information to estimate a more stable active channel design. Field surveys of the existing channel combined with the regional data produced a design of an active channel of 13.5 feet wide by 1.5 feet deep. The floodplain ranges from 6 to 20 feet wide within a 40-foot-wide corridor. No hydraulic modeling was done for this project because it was located in a park setting that could absorb high flood flows. It was clear that the channel and floodplain modifications would only result in improvements to containing flood flows and reducing the very apparent bank erosion and instability.

Owens Viani and Bradt used the design-build model and employed small family-business construction firms and conservation corps labor at a total cost of $150,000. The project removed failing concrete and riprap. It also regraded a channel and floodplain, with planned access to the channel, which is about 5 to 6 feet below the grade of the park with gentle 1-to-3 to 1-to-4 side slopes as shown in figure 3.49 (Owens Viani 2000b, 2013). Flood flows in 2001 indicated that the dimensions selected support a stable channel and floodplain as shown in figure 3.50.

A significant challenge for this project was its location in a park heavily used by young children engaged in active sports. After the fencing was taken down immediately after construction finished, the children would run down to the newly restored creek and trample the plants. Plants would be pulled up. The city irrigation system over- or underwatered plants and periodically fell into disrepair. The project sponsors eventually realized the need to hire an around-the-clock guard for the site to protect equipment and the project area during construction and planting.

The project designers experienced extensive problems with maintaining an

FIGURE 3.48 In 1999, before the project, Baxter Creek in Booker T. Anderson Park was a bare dirt channel with failed concrete (top) and rock riprap (bottom) banks. *Photo credit: Lisa Owens Viani.*

FIGURE 3.49 Baxter Creek at Anderson Park was regraded to equilibrium dimensions and meandered under existing large trees, as shown here in 2000. *Photo credit: Lisa Owens Viani.*

effective cofferdam to keep water off the construction site, and this aspect of the project became a major expense. Weed invasion became a major issue, and this factor, along with the trampled container stock, led the project managers to understand that they needed to observe the few species that were surviving well in the challenging conditions at this site and concentrate on continuing to plant them. These species included both red and white alders, cottonwood, and willow. Toyon planted as container stock thrived and also appeared on the site through volunteer establishment, but many of the common riparian shrub species failed to thrive. The project sponsors wanted to fence the site off for a longer period to allow for better establishment of the plants, but one of the funding agencies did not want fencing on the site because they did not want to appear to be cordoning the site off from the public. This policy frustrated attainment of restoration objectives and added time, grief, and expense to attempt to keep up with the ongoing damages to plants. Despite these frustrations, the trees and enough shrubs had fully colonized the site by 2004 as shown in figure 3.51, and weed species had become a minor concern.

The project designers also learned about the importance of a wide floodplain

FIGURE 3.50 Baxter Creek at Anderson Park remained stable in the winter flood flows of 2001. *Photo credit: Lisa Owens Viani.*

to absorb the backwater from the downstream culvert. The uniform 1.4 sinuosity constructed at the site was readjusted by the creek, increasing the design sinuosity at either end of the project while reducing it some in the center. The starting plan-form enabled this self-adjustment to succeed without excessive erosion (Owens Viani 2013).

Owens Viani's environmental objective was to create bird habitat, and she directed birding trips to the site for Stege School students. Her bird list as early as 2000 included phoebe, cedar waxwing, cliff swallow, downy woodpecker, great egret, and several species of warblers and hawks (Owens Viani 2000b).

FIGURE 3.51 By 2004, the Baxter Creek canopy was closing over the channel. *Photo credit: Lisa Owens Viani.*

Maintaining Riparian Corridors in High-Crime Areas

The pastoral peace of Baxter Creek at Booker T. Anderson Park was broken in the late fall of 2008 with what some of the city employees now refer to as the chainsaw massacre. Members of the Friends of Baxter Creek converged on the site in late December to add their live Christmas trees to the riparian corridor, only to find trees with their tops cut off and branches chainsawed from the trunks as shown in figure 3.52. The landscape was unrecognizable from its previous form. Calls to the city revealed that a terrible mistake had occurred. Tony Norris, director of the parks department who was part of the project's history, was on leave that week, and during that time, a call had come in from a neighbor. She complained that she wanted to see through the riparian corridor to the park beyond it and requested that the creek vegetation be cut down. Uninformed about the history of community involvement with the project, maintenance crews dutifully showed up and put their chain saws to work.

A meeting of stakeholders was organized by very apologetic City of Richmond officials, who included neighborhood representatives, the original project sponsors, members of the city council, and state regulatory agencies to pledge that they would restore the riparian corridor. A nonprofit organization was given the task to work with the neighborhood and replant the site. A neighborhood meeting

FIGURE 3.52 In 2008, the Baxter Creek project sustained an unplanned chainsawing of the riparian vegetation.

and walk-through of the site revealed different perspectives on how the corridor should be managed. Some people were concerned with the safety issues inherent in a neighborhood known for crime and believed that a creek corridor with vegetation conflicted with safety. Others did not see the corridor as a cause of crime and identified the issue as needing to manage crime, not the vegetation. This latter group tended to include those who were closer to the youth programs that had been involved in planting and maintaining the site over time. A middle ground was negotiated to provide three sight corridors through the area so that the staff in the community center could see over to a playground area located on the other side of the creek.

By the fall of 2009, the nonprofit organization tasked with the revegetation produced a replanting list that was reviewed by the original project designers and the regulatory agencies. The criterion most affecting this plant list was the desire to plant low herbaceous plans that would not affect the view corridor and would be "attractive" plants that people would relate to. The list had the classic, now often experienced problem of emphasizing popular landscaping plants that typically occupy chaparral environments as opposed to riparian environments. The riparian species were selected on the basis of restricting the list to low-growing plants, but from experience with other projects, many of them were known by restoration practitioners to be high risk for low survival as container stock in riparian corridors, particularly for an area that was now greatly disturbed and open to the sun. Input from the original designers on what species had thrived on the site in the past had little influence on the plant list. Another important emphasis of the new planting design was to develop bee habitat. An attractive landscaping of chaparral plants was designed and installed at one of the entry points to the creek corridor to make the whole site more visually inviting.

The new planting plan was installed in 2009, and a follow-up visit in 2010 revealed a site that was not focused on the objective to recover the site from the disturbance caused by the vegetation removal. The plants were selected to be low, herbaceous, and attractive, but they had not functioned to help prevent a weed invasion or to check erosion. The site was a bed for weed invasions created by opening the canopy to substantial sun, loss of cover, and little or no replacement of the cover lost.

The City of Richmond staff decided to take the lead on the project. Lynne Scarpa, representing the city Public Works Department, and Mark Maltagliata, representing the city parks department, teamed up to oversee the removal of weeds and replant the site. The invading thistles and other weed species were pulled out, and they trucked in massive amounts of mulch, providing about 1 foot of cover on large portions of the site. The city replacement plantings used species known to have high survival rates in other East Bay urban stream projects as well as good previous survival at the site. They emphasized toyon, wild rose, dogwood, cottonwood, alder, oak, and, for well-shaded areas, snowberry. The city arborist oversaw the planting and monitored plant success. The water board sponsored an on-site workshop in soil bioengineering for the city staff who sheepishly identified themselves as "chainsaw massacre" participants. The workshop empowered the city employees by adding an important tool to their management strategies, and it was clear that they were invested in the sensitive and respectful stewardship of the environment at the site. Almost-annual plantings and soil bioengineering projects have been conducted at the site for the city by volunteers from the East Bay State University environmental sciences class, the Watershed Project, and the California Urban Streams Partnership.

A monitoring report conducted by the city arborist on the site in 2011 to assess the 2009 planting indicated that there was 0 percent survival of the bee pollinator species planted with the one exception of scrophularia, or bee plant, which did very well. (An irony is that the willow species not favored by the 2009 landscape designer comes in second in a list of 456 species that help sustain desirable insects, including bees; Robbins 2013.) Very low survival rates were achieved by chaparral species, between 0 and 14 percent (with one exception, which had a 39% survival rate); a low 15 percent survival rate was achieved for ferns; and high survival rates were achieved for oaks, ninebark (*Physocarpus capitum*), coffeeberry, and wild rose. Additional plantings by the city of alder, cottonwood, maple, toyon, coffeeberry, dogwood, and snowberry thrived. The added dogwood, snowberry, and ribes species as well as thimbleberry had high survival rates if placed under the canopy shade. A June 2014 field trip indicated that the site had recovered from the 2008 event, with the return of the canopy and a thriving of these select shrub species in accordance with the experience at other East Bay restoration sites. The wishes of the original project designers, Owens Viani and Bradt, to provide urban bird habitat were strikingly realized as I saw a barn owl fly a few feet in front of me as it returned to its nest in the canopy nesting box.

In a 2014 visit, the attractive landscaping project with flowering chaparral and drought-tolerant species had mostly devolved to a tangle of weeds with invasive nonnative blackberry and two wildflower species hanging on. The lesson on the attractive gardening projects at this site, of which there have been many conducted by different nonprofit organizations working with young students over a number of years, is that they cannot be sustained unless there is a regularly hired gardener to maintain the landscaping.

In June 2014, I received a call at my government office from Richmond's chief of police about Baxter Creek at Booker T. Anderson Park. He asked me to meet with him and the police captain to talk about an incident in which a man who had wandered into the park from a newly settled homeless encampment located under the nearby freeway had jumped out of the trees at a teenage girl riding her bike along the trail next to the creek. The girl ran away physically unharmed and got help, and the man was apprehended by the police. Understandably, this incident caused concern. A woman from a neighborhood association located elsewhere in the city demanded that the police department cut down the riparian corridor. Scarpa of the city's public works department, one of the park managers, swung into action and immediately called community members from the neighborhood together. They discussed the need to reopen some of the sight lines agreed to in 2009 that had started to grow in. The police assured Scarpa and me that they understood that trees do not cause crime and that they would not support harming the riparian corridor because doing so would not rationally address the problems associated with the homeless encampment.

Scarpa, representatives of the police department, parks department, and California Department of Fish and Wildlife, and I walked the site with a mix of people who arrived bringing their range of biases for and against the creek vegetation. The different opinions expressed on this group walk provided a consensus vegetation management scheme employing the opening of the sight lines. The city staff is now very knowledgeable and experienced and have carried out the management of the corridor, which entails adding more canopy species, removing dead limbs, and helping the canopy trees flourish while protecting the shrubs as part of the multilayered mix. Figure 3.53 shows the Baxter Creek environment that they have created, with midstory and canopy trees. Figure 3.54 shows the managed "windows" through the riparian corridor, providing the desired sight lines to the child play areas from the park community center.

FIGURE 3.53 The Baxter Creek riparian recovery work has re-created a riparian corridor with an understory, midstory, and canopy forest.

FIGURE 3.54 In 2014, a view corridor project was completed that protected the riparian habitat while providing "windows" through the corridor at key locations for sight lines.

Baxter Creek Gateway Project, El Cerrito, 2005

Location: San Pablo Avenue and Key Route, and Conlon Street, El Cerrito, California
Drainage area to project site: 0.71 square mile
Project length: 750-foot valley length, 950-foot channel length
Park acreage: 1.64 acres

Project History

Handwritten notes distributed to neighbors' doors on December 28, 1992, said, "If you are interested in the possibility of turning the ugly, garbage strewn field beside the Lucky's store on Key Route into a park with trees and a clean creek running through it, please meet at the field on Sunday." This grassroots effort by the newly forming Friends of Baxter Creek culminated with a city-supported creek restoration and green space "gateway" to the city in 2005. Without this group, the creek would have remained an unnamed watershed on the city and county maps. As is often the case, the realization of this vision required a long and sustained community effort (Friends of Baxter Creek 1998a). Indeed, the strategy to build momentum through the project at Booker T. Anderson Park in 2000 helped build support for this project.

The setting for the Baxter Creek Gateway project is a 1.64-acre property in a commercial zone along the major state regional road, San Pablo Avenue, which connects East Bay cities. Occupying the northern end of the parcel was chain grocery store that was struggling financially and was considering enhancing the field of weeds next to it to improve its business and expand its loading dock. An Atchinson–Topeka–Santa Fe (ATSF) railroad right-of-way runs through the site. The rail company sold a portion of the right-of-way to BART for its elevated rail system but retained the rest of the property. The ATSF was interested in selling the remaining portion of the property to the City of El Cerrito redevelopment agency. A regional trail tying the cities of Berkeley, Albany, El Cerrito, and Richmond was being developed along the BART corridor and ended here. The Friends of the Ohlone Greenway formed around this site to plant trees and improve this part of the city, and the Friends of Baxter Creek grew from this effort. The vision of the group was to restore the creek on this site through what was later called the Gateway project; extend the trail across San Pablo Avenue to another vacant lot, later known as the Ohlone Gap site; and continue the creek restoration along the trail at this site.

The Friends of Baxter Creek encountered barriers within the city government, which took the position that the site was a ditch of little environmental value. The city was more interested in development opportunities. Baxter Creek would periodically overflow into San Pablo Avenue, and city crews would bulldoze the creek on these occasions, vainly attempting to stop the flood damages. The city's plans for the site were to fill in the undeveloped acreage with an auto supply store and other big-box stores (Owens Viani 2013). The efforts by the Friends of Baxter Creek to protect the greenway and creek for a future greenbelt and trail required constant vigilance and pressure. Letters to members of the city council and appearances at their meetings raised the issues that adding a fence to the gateway site to keep the public out was turning it into a dumping ground; that a new Honda dealership occupied the Ohlone Gap site across the street and cars were being parked on the creek bank; and that the Adachi Associates property owner of the Ohlone Gap site tried to strike a deal (apparently for $3,000) with the City of Richmond to reroute any trail off their site onto a street to the south. This rerouting would have put the trail at a dead end at Interstate 880 and prevented the extension of the trail into Richmond. The efforts to stop extension of the trail involved unspoken desires among a number of entities to not connect the trail to what was perceived at the time as a less-than-desirable community (Friends of Baxter Creek 1998b).

By 1998, the fate of this parcel came to a head when a Lucky's store redevelopment proposal involved filling in part of the creek. Citizens opposed the channel filling but met resistance at city council meetings for the park restoration plan (Friends of Baxter Creek 1998c). Ultimately, the collapse of the Lucky's grocery chain reopened the future of the land. The Friends of Baxter Creek reminded the City of El Cerrito that the City of Richmond had just reaped the benefits of its project at Booker T. Anderson Park.

The Friends of Baxter Creek (FOBC) essentially tapped into an existing, informal group called the Friends of the Ohlone Greenway, which sponsored tree planting and community awareness projects. This group included Lisa Owens Viani, neighbors, and Jeff Cruzan, a university graduate student in geology who literally put Baxter Creek on the map by developing a sketch that showed the Ohlone site with a creek. Later, Steve Price, an urban ecologist active in the Friends of the Ohlone Greenway who worked for the new urbanist Peter Calthorpe, put together a vision plan with photos and graphics to produce a concept plan for this site and the vacant lot across the street, the Ohlone Gap property. Owens Viani of FOBC, now an experienced grant writer, along with Maryann Aberg, an El Cerrito resident, approached the California Coastal Conservancy for funding to acquire the parcel on the west side of San Pablo Avenue, El Cerrito, from the railroad, primarily to serve as a much-needed urban open space project in lieu of the dense development proposed. They expanded their vision by adding a trail to acquire support from the bicycle community, and creek restoration became a focus based on Owens Viani's acquired experience.

This fundraising strategy and land acquisition succeeded in 2002, and by 2003 gave the City of El Cerrito a new opportunity for a green space. The California Coastal Conservancy grant of $350,000 added to the city's contribution of $97,400. Another critical development at this time was changes in El Cerrito public works staff. The support of Carly Payne and Melanie Mintz became central to the city's shift in course, and they set about arranging for a planning process for the site. It was also critical that Tom Butt, a member of the Richmond City Council, was supporting the whole vision plan with creek restoration for the site (Owens Viani 2000b, 2013). The unrelenting perseverance of the Friends of Baxter Creek—by 1998, a five-hundred-member organization—finally turned the political opposition in to support.

Project Planning, Design, and Construction

The site acquired a new name at this juncture, the Gateway Park, and the so-called ditch was given the name Baxter Creek. These names recognizing the value of the resource are important for steering the future of urban parcels. In May 2003, the City of El Cerrito, Friends of Baxter Creek, and the Aquatic Outreach Institute collaborated on organizing community meetings for planning the future of the site. They secured the services of the National Park Service River, Trails, and Conservation Assistance Program to sponsor a professional, public-based design process. The city took its role of creating an open, transparent, and inclusive planning process very seriously. The planning was informed by a previous truncated planning process on another creek project in the city that had gone terribly wrong, with a project being installed without citizen participation or the proper review and permits. The city council created a design review group, and by 2004, it hired the Restoration Design Group (RDG) to sponsor follow-up design charrettes with the interested public. RDG provided slideshows with some project site options and supplied ribbons and design-kit pieces so that the workshop participants could build physical models of the site. Forty to fifty people gathered in teams to explore design concepts together. The planning process had to contend with the negative history of the site, which involved homeless encampments, lots of trash, and the ditch that was the creek (fig. 3.55). Both the police and fire departments were part of the design team (Goetting 2013; Mintz 2014).

In 2004 and 2005, the restoration design and construction plans received support from the State Water Resources Control Board in the form of a CALFED grant of $492,000, with an additional $100,000 provided by the California Coastal Conservancy. The city's manager worked with the city redevelopment agency to secure $288,000 for trail development, lighting, a San Pablo Avenue entry and seating area, and interpretive displays. The multiple public design meetings secured good public support for the project. Bob Birkeland, the landscape architect

FIGURE 3.55 Before it was restored as part of a new urban greenscape, Baxter Creek was a ditch in an abandoned railway right-of-way located on a major commercial route.

from RDG, noted how the feedback of the site's current users involved in tree plantings and other stewardship activities influenced the final design plans (Mintz 2014).

RDG completed construction in 2005 (fig. 3.56). The project design relied on hydraulic geometry relations and regional curve information to design channel dimensions and the meander belt. The creek stability has been achieved with equilibrium channel dimensions and soil bioengineering with little or no use of

FIGURE 3.56 Project construction at Baxter Creek transformed derelict space (above) into a gateway to El Cerrito (above, opposite page), featuring creek restoration, a regional trail, and public congregation areas. *Credit: Chris Benton.*

rock. RDG reused excavated soil from the creek channel and floodplain development to grade an interesting and varied topography along the creek corridor and
trail. A design lesson from this project was that weed control was a major issue.
Although a native seed mix was used to cover the site, in hindsight the designer
believed that selecting one resilient species of grass would have been more effective. Areas exposed to the open portions of the site outside the riparian corridor
were particularly vulnerable to weeds, and the city contracted with the nonprofit
Aquatic Outreach Institute for a number of years to organize volunteers to do
hand weeding. Ultimately, some of this weeding resulted in accidental removal
of intended natives, and the ground cover became nonnative within a few years.
The trees have been the most successful plantings on the site, and a more concentrated use of larger container stock would have been an advantage (Goetting 2013;
Mintz 2014).

Beginning in 2013, the site came under the supervision of the city's watershed
and landscape manager, Stephen Pree. The Gateway Park is a head turner for
people driving and biking along this corridor, with the urban concrete suddenly
transformed into a green oasis. Flooding the streets has not been an issue since the
project was installed, and the new project was tested by high flows in 2006 (Mintz
2014; Ortiz 2014). Interpretive panels installed by the city show the Baxter Creek
watershed and explain how a series of projects along the creek are now "beads on
a green necklace" winding through the city (Mintz 2014).

Ohlone Gap Green Way, Richmond

Location: San Pablo Avenue, near MacDonald Avenue, across the street from the
Gateway project, Richmond, California
Expected construction: 2016
Drainage area to project site: 0.71 square mile
Project: 400-foot creek valley length, 900-foot channel length
Park acreage: 1.75 acres

Opposite the Gateway site in El Cerrito is the vacant parcel of land in adjacent
Richmond that was in the sights of the original Friends of Baxter Creek vision for
the San Pablo Avenue corridor. The owner was the Adachi Nursery. The nursery
operated until the early 1990s, the city acquired it after it closed, and a decade
later, the site was being considered for various development proposals. The BART
crosses San Pablo Avenue to this property, so it was logical to continue the greenway concept on this parcel. The Friends of Baxter Creek asked RDG to provide a
concept stream restoration component of the greenway plan for the parcel. The

Friends used this plan to compete with the senior housing development plan proposed for the site. City councilman Tom Butt was instrumental in securing the site for a different land use than housing, advocating open space and creek restoration and an employment center dedicated to creating jobs, a popular political cause in Richmond. The federal financial sponsor of the senior housing concept ultimately withdrew its support because the nearby BART track noise exceeded its standards for housing.

Joe Comancho on the City of Richmond staff took advantage of a new State of California Urban Greening grant program sponsored by the state's Strategic Growth Council. His application to complete a green corridor through the cities of El Cerrito and Richmond was awarded a grant for $888,000 in 2013 because the project was considered a significant contribution to climate change readiness in that it continued a greenbelt through an "urban heat island" and continued the opportunities for regional bike transportation by further developing the trail. Another grant of $529,000 was awarded from the state's Safe Routes to Transit program. The creek design takes the experience gained from the project immediately upstream at Gateway and applies regional hydraulic geometry for channel dimensions and meander pattern. An HEC-RAS model was used to estimate water surface elevations because of a flooding history in the area, and model discharges, velocities, and dimensions were used as a cross-check on the hydraulic geometry. This project extends the trail system to Interstate 880.

Mira Flores Project, Richmond

Location: 14-acre site on the west side of Interstate 880 bound by the BART corridor and Ohio Avenue to the north, Wall Avenue on the south, and South 45th Street to the east
Drainage area: 0.72 square mile
Project: 750-foot valley length
Park–open space acreage: 4 acres

After walkers on the Ohlone trail leave the Ohlone Gap greenway site and continue west, they pass under the Interstate 880 freeway and arrive at the historic ruins of the Sakai, Endo, and Oishi family nurseries. The three nurseries specialized in long-stemmed roses and carnations. During World War II, these Japanese families were interned in Arkansas, but colleagues held the nurseries properties for them so that the families could continue to operate them upon returning after the war. In 2006, the nurseries closed due to cheap flowers flooding the market from South America (Jones 2011). A "tiny" section of creek encased in concrete was discovered by the Friends of Baxter Creek in 2000 on the nursery site. The group learned from the property owners that one section of the creek was undergrounded in a culvert in the late 1960s to make more room for greenhouses. The nearby grove of willows and the groundwater supplies pumped by the nurseries indicated the creek's presence (Owens Viani 2000a).

In 2006, the City of Richmond's redevelopment agency purchased the property from the nursery families and developed a team of housing developers to build 80 units of affordable senior apartments and 150 market-rate but heavily subsidized single-family homes, named the Mira Flores development. The project involves a layered financial structure of federal, state, and local housing funds. The expected date of construction for the housing is 2017. The development plan includes 4 acres of open space and the daylighting of Baxter Creek for 750 feet in the open space. Two historic farmhouses are to be preserved, and two or three of the greenhouses will remain, with the potential to be leased to urban gardeners (Jones 2011). The Great Recession and housing market crash delayed the development start, but the City of Richmond decided to move forward first with the development of the green space. The Richmond city staff applied again to the state's Urban Greening grant program for $1,664,319 to move forward with the park development and creek restoration, which begins in 2016–2017. The site plan was developed by Gonzales Architects and Patillo Garret Associates Design, and the concept creek restoration plan was developed by RDG. Hazardous waste, including pesticides and lead, has been removed through excavation of the top layers of soil. In June 2015, the California Strategic Growth Council awarded a $5.1 million grant for Mira Flores senior housing construction for 80 affordable homes for low-income seniors. Because the project features proximity to a mass-transit BART station, shopping center, and greenway, the project qualified for California's climate change Greenhouse Gas Reduction Fund (Goetting 2013; Scarpa 2014; Butt 2015).

Village Creek Daylighting in Housing Redevelopment, Albany, 1998

Location: Jackson Street at the southern boundary of Albany Elementary School at the site of old Riley Street, University Village, University of California, Berkeley family student housing, Albany, California
Drainage area to project site: 0.15 square mile
Original project: 900-foot valley length, 1,440-foot channel length
Final project: 700-foot valley length, 1,120-foot channel length

Project History

The location for the Village Creek daylighting project and an adjacent Codornices Creek project (fig. 3.57) has a fascinating and unique history. The project takes place on a 62-acre site that was originally a bayside wetland/floodplain, but it was transformed into an instant community of workers for the World War II Kaiser shipyard in Richmond and naval personnel housing. An aerial photo from 1947 (fig. 3.58) shows the conversion of an equestrian racetrack at the top of the photo into the Naval Landing Force Equipment Depot set up in 1944 to reinforce the West Coast from a feared Japanese invasion. The 3,000 naval depot personnel desperately needed housing, and many of them were set up in housing provided by the Federal Public Housing Authority on the wetlands and floodplains of Village and Codornices Creeks (Lee and Lee 2000). Most of this acreage was in Albany, but some of the housing extended across Codornices Creek into Berkeley.

The shipyard workers needed a way to go north around the bay to reach the Richmond shipyards, and as seen in figure 3.58, the center of the 1947 photo shows a wide, curving feature that contained the tracks for the Richmond Shipyard Railway. Abandoned parts and rails from San Francisco, Napa, Los Angeles, and even New York transit systems were quickly cobbled together to provide this transit line, which was referred to as a "make-do" line. The federal public housing project was called Codornices Village after the stream that can be seen flowing through the left side of the photo in figure 3.58. The creek follows the tree line from the photo's left center that disappears into a large, black, long, L-shaped building toward the upper left quarter of the photo. Here, Codornices Creek flows into the Steel Tank and Pipe Company, which occupied the site starting in 1923 and which was sold to U.S. Steel in 1950. Village Creek is very visible in the photo to the right center. A dark, accordion-shaped line of trees can be seen flowing from San Pablo Avenue from the bottom of the photo in the direction of the racetrack at the bay. (The top of the photo is west.) The creek ends at the curve in the Richmond Shipyard Railway. A 1946 aerial photo shows a hook-shaped connection with the rail line, which was a pile-bent bridge crossing over the Southern Pacific rail line to send the workers to Richmond. This bridge was removed by the time the 1947 photo of figure 3.58 was taken.

David Kinkead, a federal public housing manager, was a pioneer in developing integrated public housing and equal employment opportunities at Codornices Village. Another photo from 1947 shows that the employees of the Codornices Village are approximately half men and half women and that about one-third are white, one-third are Asian, and one-third are African American (University of California, Berkeley 1997; Lee and Lee 2000).

Village Creek is actually the remaining aboveground remnant of Marin Creek, a moderate-size watershed for the East Bay that was culverted so that Marin Avenue could be constructed on top. This remnant creek now has its own separate watershed from the culverted Marin Creek, hence its relatively small size for being located at the bottom of a watershed near the bay. By the 1990s, when I first visited the site where Village Creek was eventually restored the creek was located underground near Riley Street as shown in figure 3.59.

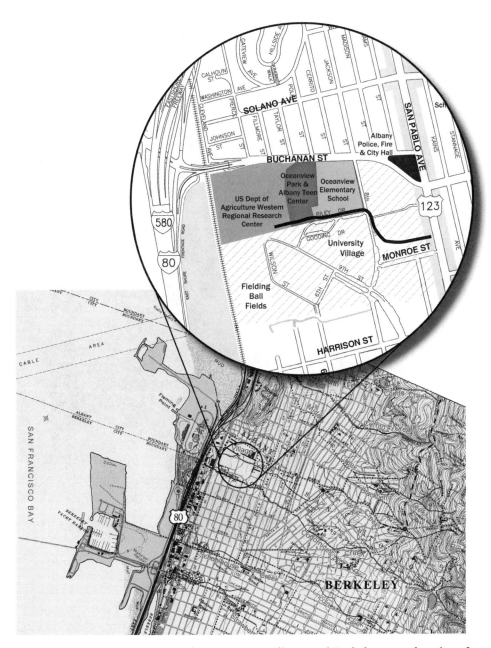

FIGURE 3.57 Village Creek is located in western Albany and Berkeley, near the edge of San Francisco Bay. *Credit: Lisa Kreishok.*

FIGURE 3.58 A 1947 aerial photo shows the World War II military and shipyard workers' housing, Village Creek, nearby Codornices Creek, and San Francisco Bay.

In 1956, the University of California, Berkeley purchased the acreage and World War II housing from the federal housing authority at a bargain price of $55,000 and renamed it University Village. The university built another 500 units in the early 1960s, creating a mix of 1940s and 1960s housing units. Although the federal housing was built to be temporary housing, the university used this housing stock until it began to implement a redevelopment project to replace it with modern units toward the late 1990s. The university-added buildings brought the population to 2,700 residents in 920 units (University of California, Berkeley 1997; Lee and Lee 2000). Master planning for upgrading the area began in 1993. Both the Village Creek and later Codornices Creek restoration projects began as

FIGURE 3.59 Before the restoration project, Village Creek flowed in an underground culvert immediately south of Riley Street.

a result of the redevelopment plan, which involves a phased removal of the old housing stock and 920 replacement units. Much of the phased redevelopment had been completed by 2014, but some upstream housing units are still to be constructed.

Village Creek used to flow to the south of Riley Drive, near Gooding Drive. (Riley Drive is named after a beloved fire chief in Albany in 1944 and is not related to the author.) A University of California landscape architecture class submitted a design to the Campus Planning Office that would integrate restoring the creek through the new housing development as a feature in a public plaza and create a congregating space for the residents (Isbill 1988). A tragedy of the Village Creek project is that the basic principles of good urban design and creative suggestions from within the university were ignored; instead, a grid of housing and parking lots was imposed on the landscape, without the natural or social needs considered by the university architectural and development office.

Ultimately, the motivation of the University of California to sponsor a restoration project for Village Creek was to provide for the acres of mitigation required for wetland impacts from its housing redevelopment and to compensate for relocating the creek and its culvert to the northern boundary of its property to get the creek out of its development envelope.

Project Design and Construction

In the late 1990s, the Planning, Design and Construction Unit of the University of California, Berkeley began planning for a Village Creek project in a section of the 1940s housing. By this time, Village Creek remained only as a ditch west of the project site, along the US Department of Agriculture research building where it joined a bypass channel for Codornices Creek to enter a culvert under Interstate 880. An open 700-foot segment also remained below San Pablo Avenue to the east of the project site. The initial objective was to bring the creek to the surface, but the planners wanted to minimize the space that the creek would take from the development. They hired a environmental engineering firm to design a straight creek channel along the perimeter of the development on the border with the Albany Elementary School. The firm obliged with a trapezoidal channel within a 23-foot-wide corridor. The trapezoidal channel was designed with a channel slope of about 0.01, a roughness of 0.045, design flood of 207 cfs and 4.8 feet per second velocities, about 1.5 feet of freeboard for the one-hundred-year flood, and 2-to-1 side slopes up to the ground level. The design active channel was 3 feet wide and 2 feet deep, for 6 square feet of cross-sectional area. Hydrology was computed using both an HEC-1 hydrology model and rational method hydraulic analysis (Jones 1998). The ten-year recurrence interval discharge was estimated between 80 and 134 cfs, the ninety-five-year discharge was estimated at 95 cfs, and the one-hundred-year discharge was estimated at 207 cfs.

The university and the City of Albany planners developed a relationship with the nonprofit WRI, which had restored a problematic section of Codornices Creek along the Ehret Plumbing building in 1995 and had helped the property owners avoid flood damages in the recently experienced twenty-five-year flood. This project created goodwill among the adjacent property owners for stabilizing an undersized creek channel by adding proper dimensions and soil bioengineering. The WRI had also helped stabilize an undersized channel at Eighth Street on Codornices Creek in 1997 and 1998. The agencies also noticed the low price tag using a nonprofit organization with a restoration approach. The WRI was invited to review the above design of the engineering firm.

The WRI used reference information from the open channel immediately upstream and information from the regional curve to determine that the cross-sectional area needed for a stable channel was about 5 square feet, a similar active channel design as proposed by the consulting firm. The WRI, however, wanted to provide for a more stable system by adding a floodplain, channel sinuosity, and riparian corridor that did not need vegetation removal for flood maintenance. A sinuosity was selected on the basis of conditions from the 1947 aerial photos, and a floodplain of 15 to 18 feet was provided to accommodate the meander belt. A roughness of 0.075 was assumed for the design so that the university could avoid costly ongoing maintenance to achieve the proposed 0.045 roughness value. The WRI also recommended a sustainable riparian habitat with water quality benefits that could qualify as a mitigation project. The meander belt design was based on hydraulic relations between channel width and length and the average radius of curvature and amplitude dimensions for the width, which indicated that a project right-of-way of 30 to 32 feet would be needed to meet flood control objectives and provide a stable planform site. The WRI design increased the entire project area from 33 square feet to 81 square feet. The developer consulting firm, J. R. Roberts Corporation, confirmed through an HEC-2 model run used to determine flood elevations that these design dimensions would meet the flood protection objections. This design only increased the width of the corridor by 6 to 8 feet. Because this corridor width difference was insignificant and the corridor had more mitigation value, the university proceeded with the wider WRI design concept.

What was the basis for the design? The design considered a 1865 coastal survey map that showed both Codornices and Village Creeks as single–thread channels that disperse most likely as discontinuous channels as they enter the bay, although any details of the lowermost sections is not captured by the survey (US Coast Survey 1865). A 1930s aerial photo gives a better view of the relatively wide riparian corridor along the single-thread channels of both Codornices and Village Creeks until they dissipate into the bay near the railroad tracks. The 1947 photo (fig. 3.58) indicates that the location on which we were to work historically had a single-thread channel that was remarkably sinuous within the confines of the urbanizing environment. The channel appears like channels described in some geology or fluvial geomorphology books as an "underfit" channel in which a meander pattern is controlled by adjacent constraining terraces (Richards 1982). It seemed reasonable to emulate this pattern again at the same location; if the creek were designed with too much sinuosity, it could cut off meanders or change the meander-to-channel-length ratio as it evolved. The photos show that the riparian forest appeared to be lush between the 1930s and 1940s and that their locations may have been influenced by agricultural activities, with a channel flanking each side of an agricultural field. For lack of any other analog, we decided to select the 1947 conditions. The upstream remaining above ground 700 feet of Village Creek was in stable condition and provided a reference for a bankfull cross-sectional area of 4 square feet. This channel is located on a steeper slope and was not considered a good reference for sinuosity at about 1.3.

We made a photographic slide of the 1947 print photo (fig. 3.58) and projected it on a white wall with a slide projector. Tracing the meander pattern indicated a sinuosity from 1.8 to 2.0, which seemed to be an unusual planform for an East Bay creek entering the bay. Should we use this value as a reference condition? The

1930s photo confirmed the relatively wide riparian corridor on a straight trajectory to the bay, but it did not have the resolution to represent the channel sinuosity. We would be continuing the narrow valley corridor in place since the 1930s and had no choice about this aspect. We proceeded with the high sinuosity concept to see if this flat planform channel type could represent a sustainable, relatively stable planform in this situation.

To create a random meander on the plan, I took a premeasured length of string to represent channel length, placed it on the plan, closed my eyes, and arranged it on a selected valley distance on the paper to represent the design sinuosity. After drawing in the random meander, I modified the radius of curvature and amplitude for some of the bends I made so that they fell within reasonable ranges of values for average hydraulic geometry relations with channel widths and wavelengths. The highest I could push the sinuosity within this constrained space was 1.6, and the final design channel slope was 0.005. The design plan in figure 3.60 illustrates how a meandering channel was fit into a straight corridor.

The Village Creek was a design-build project with the WRI teaming with Dick Botkin and Jim Sly of J. R. Roberts Corporation. Josh Bradt and Mike Vukman of the Urban Creeks Council also supervised construction and plant installation. The channel layout survey required one day; excavation took four days, with soil reused at other locations on the project for housing platform fill. The combined J. R. Roberts and WRI costs—including design, permitting, irrigation system purchase and installation, plant collection, plant purchases, conservation corps labor, and supervision of grading—was $21,548 plus four days of equipment use and operator time from J. R. Roberts for a total estimated cost of $30,000.

Project installation issues included the need to keep costs low, so a limited plant palette was used. The use of willow posts and alder tree planting became the main framework for achieving quick cover and stream planform stability. Erosion-control fabric was limited to the active channel area, and the rest of the erosion control was accomplished by spreading straw about 6 inches deep through the site. The site was extremely compacted because of its location under a previous housing development. Gas augers had to be used to drill through compacted clay soils to plant the willow posts. We considered purchasing mycorrhizae to reinoculate the soil with microbes, which help plants convert nutrients to help support plant growth—but instead we collected willow stock and soaked the material in vats of water for a week and even more to help the material take up water and remain fresh. The plant collection opportunity and planting schedule did not line up well, so the willow received more soaking than we had planned and stretched into several weeks. Based on field surveys, the creek planform was spray-painted on the ground, which guided the excavation of the active channel within the trapezoidal right-of-way as shown in figure 3.61. Figure 3.62 shows the completed excavation

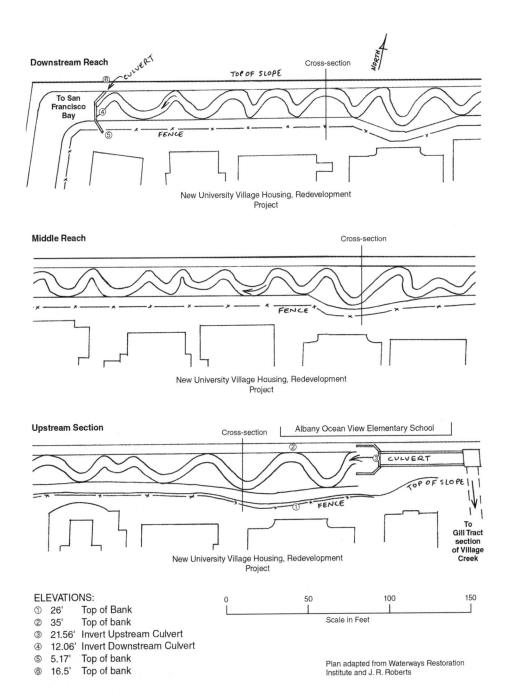

Downstream Reach
CULVERT
TOP OF SLOPE
Cross-section
NORTH

To San
Francisco
Bay
⑥
④
⑤
FENCE

New University Village Housing, Redevelopment
Project

Middle Reach
Cross-section

FENCE

New University Village Housing, Redevelopment
Project

Upstream Section
Cross-section
Albany Ocean View Elementary School
②
③ CULVERT
TOP OF SLOPE
① FENCE

To
Gill Tract
section
of Village
Creek

New University Village Housing, Redevelopment
Project

ELEVATIONS:
① 26'   Top of Bank
② 35'   Top of bank
③ 21.56' Invert Upstream Culvert
④ 12.06' Invert Downstream Culvert
⑤ 5.17'   Top of bank
⑥ 16.5'   Top of bank

0   50   100   150
Scale in Feet

Plan adapted from Waterways Restoration
Institute and J. R. Roberts

FIGURE 3.60 The Village Creek restoration design set a sinuous active channel and floodplain within a straight, trapezoidal flood control channel. *Credit: Lisa Kreishok.*

with the willow posts installed. Some conflicts arose between an excavator operator used to standard building construction and the excavation supervisor. One day I received a call from Josh Bradt, who was afraid that things were going to come to blows because the equipment operator thought that he was being made fun of with the request to excavate a very sinuous path that was the Village Creek channel.

I returned to the site after a week's vacation to start planning the revegetation of the completed excavation and found that the university had decided to fill in the last 200 feet of the stream channel. That explains the ultimate remaining channel project length of 700 feet as opposed to the published 900-foot design plan. A few weeks later, regulatory staff from the San Francisco Bay Regional Water Quality Control Board checking on the project discovered the channel filling, a violation of the permit they had issued. The university also violated a buffer zone requirement in the permit in which no structure was to be within 40 feet of the creek bank. As a result, the water board prevented the university from using the project as a mitigation site, required the submittal of discharge reports and a series of new stormwater treatment controls, and requested that the university develop a public process for planning the upcoming Codornices Creek project, which would need to be integrated into the redevelopment plan. It was reported that the university

FIGURE 3.61 An active channel and floodplain were excavated in Village Creek's trapezoidal corridor.

FIGURE 3.62 As part of the Village Creek project, installation of willow posts was the primary strategy for providing an inexpensive, stable, and shaded channel.

project manager was encouraged to leave, and that person exited shortly after (Covina 2000; San Francisco Bay Regional Water Quality Control Board 2000; Lichten 2014).

Project Lessons and Significance

The significance of the Village Creek project is that it provided a model, both in planning process and design, for the next, much larger project to occur later on the adjacent Codornices Creek. The Village Creek design was, frankly, an experiment to see if a channel with a great deal of sinuosity could provide planform stability within the terraces of a straight trapezoidal corridor. Monitoring by the WRI found that the sinuosity appeared stable for the first years of the project. Based on that result, the design approach of employing a high sinuosity channel on stream channel gradients under 1 percent in the old marsh areas was applied to Codornices Creek, located about 1,300 feet to the south. The design process involved a rethinking of what client-consultant relationships should be and added the role of educator to the consultant's job.

DESIGN PROCESS: YOU ONLY GET WHAT YOU ASK FOR

The learning experience of "You only get what you ask for" has become the guiding principle and common thread among all the cases described in this book. I came to understand that roles of the design professional were to interact with clients, offer our professional knowledge, and become an advisor and teacher for our clients to help them identify the array of possibilities for a project that they would not be prepared to identify on their own. The design process involved interacting with the client to suggest a new project right-of-way rather than assuming that our job was to simply accept the project parameters that we were given. The first consultant performed the standard way, which was to oblige the constraint given by the client on the available footage for right-of-way, and therefore developed a simple trapezoidal flood control channel with a straight low-flow channel. No inquiries were made about other possibilities for project corridor width in this redevelopment site. The WRI approach was to first design a channel estimating a stable meandering planform, with a floodplain and active channel, and substantial riparian corridor as we had seen in the historic photos. After completing this concept design, the WRI calculated the right-of-way requirements. The design was granted the additional space requested because it demonstrated a project with better performance.

PROJECT DESIGN LESSONS

What happened to those poor plants placed in soils that seemed almost like concrete? By the winter of 1999, the willow plants began to sprout, and by 2001, the creek vegetation was becoming established as shown in figure 3.63. The plant growth proceeded slower than was our experience at other sites, but by 2004, five years later, the site looked much like our other restoration sites in terms of the growth and vigor. There was good performance from the primary riparian species willows and alders as shown in figure 3.64. Dennis O'Connor, an expert in plant ecology and restoration installation who has worked with the WRI, explained our unintended great success. Willows tend to have shallow roots, especially on shallow clay-capped alluvial soil, so these willows were in an environment that they can thrive in. Dense, compacted soils can retain soil moisture longer through the dry summer months, creating a favorable willow environment. Because willow naturally attracts and supports the mycorrhizae with their roots, it is not a good use of funding to try to add more (O'Connor 2014).

Currently, a dense riparian corridor borders this new housing development. The camera-on-a-kite aerial photo of the project site in figure 3.65 looks strikingly like the creek in the 1947 photo in figure 3.58. Children from the development climb over the 3-foot-high wood fence on the south side of the creek and use this wild space as their playground.

FIGURE 3.63 The Village Creek riparian growth initially struggled under hardpan soils, but was becoming well established by 2001.

FIGURE 3.64 By 2004, there was a dense riparian forest established along Village Creek.

FIGURE 3.65 A bird's-eye view of Village Creek shows a lush creek corridor through the housing development in 2014. *Photo credit: Cris Benton.*

Plant survival at this site closely corresponded to our recent observations at the other case study sites. The planting plan relied on a short plant list of three species of willow (*Salix laevigata, S. lasiandra,* and *S. lasiolepis*); California buckeye; California bay; coastal live oak; bigleaf maple; and four shrub species: toyon, dogwood, red-flowering current, and California rose. The ground cover used was ninebark (*Physocarpus capitus*). The limited plant list was really a function of keeping down costs.

Fourteen years later, the plant survival of most of these species was quite good. In 1998, we were continuing to accumulate knowledge on which species were doing the best in East Bay creek restoration projects. This site reflected outcomes that we ultimately saw years later at other East Bay restoration sites. Exotics such as Himalayan blackberry and some Algerian ivy appeared, but the native ninebark is doing remarkably well competing with these invasive plants under the shaded canopy. The toyon performed well but is in decline because of the well-shaded corridor. No bay trees survived. By 2012, none of the red-flowering currents remained. The native shrub, coyote bush (*Baccharis pilularis*), colonized the site as volunteers. The bigleaf maples are struggling under the shade and vigor of the willows, and someone had pruned back a few willows to give the maples more light. Oak and maple trees planted on the outside of the riparian corridor are constrained by the fencing and sidewalks and are shaded by the creek-side buildings. The success of willows and alders do not appear much affected by the building shade, which suggests that if they get a good start in the sunny phase of the site and their top canopies are exposed to the sunlight for part of the day, they can prevail under these conditions. As we learned from the Baxter Creek project at Poinsett Avenue, the later-succession trees such as maples and oaks can be assisted by modest pruning to allow them more space and light.

Monitoring by the WRI in 2007, 2009, 2012, and 2014 determined that the average channel dimensions for the active channel remained close to the design, although there was a range of active channel depths from 1 to almost 3 feet and a width from 4 feet to 6 feet. Figures 3.66 and 3.67 illustrate topographic surveys performed with a level and alidade for as-built conditions, a survey by two university students in 2005 (Asher and Atapattu 2005) and a survey completed under the supervision of the author in 2014 (Logsdon and McIntee 2014). The surveys recorded that the downstream reach above the culvert was affected by backwater conditions on this very flat slope for as much as 300 feet upstream. The channel profiles over time record the lower end as depositional while the top 400 feet experienced incision from the original slope elevation. Two headcuts have been moving up through the channel, and this process has created channel reaches more than 2 feet deep in some places. After readjusting to channel clogging caused by sun exposure of the channel in its first stages and resulting growth of rushes and reeds, the downstream active channel affected by the backwater remains 1 foot deep. It shares a similar dynamic with the low-gradient Baxter Creek at Booker T. Anderson Park and Blackberry Creek in Berkeley in which the downstream portion above the culvert affected by a backwater is shallower on the average than the upstream channel reaches. Future urban stream restoration design needs to anticipate that channels upstream of culverts need to be particularly well shaded, quickly, to head off colonization by rushes and reeds, which catch sediment and produce channel filling. Over time, as the channel became well shaded at Village Creek and the cattails died, the downstream portion recovered its dimensions close to the original design.

Observations of changes in channel planform seem to indicate that the stream is flattening its slope through erosion on the outside bends of the meanders as well as headcutting. The creek adjustments may be adjustments that the stream is making to compensate for a channel slope that is too steep with a design sinuosity that is too is low. Some of the headcutting may be a response to the flashy urban hydrograph. The two adjustments taken together suggest that a more stable planform may be flatter and even more sinuous.

Observations over time suggest that the width-to-depth ratio for this channel was too high. The original design with a width of 5 feet and depth of 1 foot was

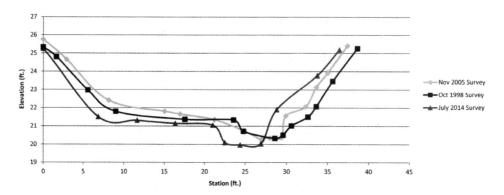

FIGURE 3.66 Village Creek cross-sectional surveys of elevation (vertical axis; in feet) at various stations (horizontal axis; in feet) indicate a range of active channel adjustments. *Credit: Lisa Kreishok.*

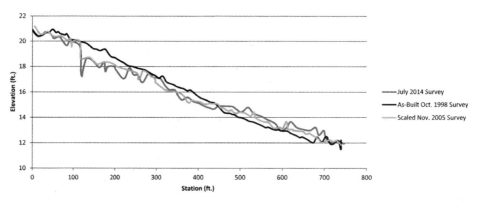

FIGURE 3.67 The profiles surveyed over time record the backwater influence caused by the downstream culvert. Sources: 1998 As-Built Survey, Waterways Restoration Institute; 2005 survey, Melissa Asher and Kaumudi Atapattu, University of California, Berkeley; 2014 survey, Logsdon Willis and Connor McIntee Watershed Stewards Project of the California Conservation Corps. (The student surveying project in 2005 used different distance stationing than the 1998 and 2014 surveys, so the profiles are matched as feasible. The cross sections were able to be well matched.)

a width-to-depth ratio of 5. In many reaches, the ratio readjusted to range from 3 to 4, meaning that the channel adjustments that occurred were greater in the depths compared to the widths. This finding suggests that we should have been more aware that this channel type is a low-gradient tidal channel once influenced by downstream bay and tidal environments (even though the tide may not currently reach this location any more) and therefore will tend to have a lower width-to-depth ratio than streams on slopes above 1 percent gradients. Cross-sectional surveys performed on other East Bay streams in similar low-gradient locations near the bay, such as Wildcat Creek and Codornices Creek, found that fluvial stream channels (as opposed to tidal channels) similarly adjusted to lower width-to-depth ratios than the channel reaches located farther upstream.

This restoration project, involving the creation of a meandering active channel and floodplain, has a perfect control site immediately downstream for a comparison to a project that did not provide sufficient floodplain, directed flows into an undersized channelized low-flow channel and used a low roughness factor for design of flood conveyance. This downstream section of channel was constructed by the university, which was no longer interested in mitigation projects, as a low cost, minimum right-of-way project to continue the Village Creek flows downstream to the bay. This channel fills with sediment because the lack of canopy trees allows sun-loving plants such as cattails and other rushes and reeds to fill the channel and collect sediment. The low-flow channel is re-creating a meander, and as it shifts, it buries the stormwater outfalls to the creek. The backwater from the downstream culvert under the railroad tracks has no floodplain for storage of the flows. The blocked stormwater outfalls do not allow local drainage. The university had to go to the considerable expense of dredging out the channel because it overflowed onto the adjacent parking lots. This dredging activity put the university into an adversarial relationship with the environmental regulatory agencies, which then required that the university plant canopy plants to shade the channel, thereby controlling the cattail growth and slowing the need for regular dredging (May 2014).

THE FAILED EXPERIMENT FOR A COLLABORATIVE UNIVERSITY FIELD PRACTICUM

Discussions with William Jordan III, the original editor of the Society for Ecological Restoration journal, *Ecological Restoration*, raised the topic that by the late 1990s, universities and colleges were not addressing the need to better integrate science and practice in restoration, noting the reticence of numerous academics to integrate restoration design practice into curricula. This reticence was occurring even as students expressed their needs to be exposed to practice as part of their education. During this time, I received a call from a well-known engineering professor at the University of California, Berkeley asking if I would set up an off-campus stream restoration practicum that could expose his students to current restoration practices in which civil engineering would be combined with other disciplines. The National Fish and Wildlife Foundation was in favor of supporting a nonprofit organization–university pilot program in which the WRI would hold an off-campus course taught by academics and practitioners to explore and evaluate the tools in use for designing and constructing restoration projects. The Village

Creek project design and construction project lined up nicely with the calendar for this pilot program.

The WRI developed a field practicum course that included the lecturers Luna Leopold, professor emeritus of the University of California, Berkeley; Peter Goodwin, who was leaving private practice to teach at the University of Idaho's engineering department; and other practitioners from Bay Area consulting firms. The two-unit course drew students from the both the Davis and Berkeley campuses of the University of California and met for two hours a week with twelve hours of field time over the course of a semester. The purpose of the course was to integrate the fields of fluvial geomorphology, hydraulic engineering, plant ecology, landscape architecture, and community involvement and organizing to develop a realistic experience for students interested in pursuing restoration practice.

The class produced a restoration design report for Wildcat Creek at Davis Park for the City of San Pablo. It also participated in the review of the Village Creek project design. The students were brought to the site to watch the construction and to take part in a hands-on class installing soil bioengineering. Many of the schools of restoration presented in chapter 2 were covered in the course material. Written student evaluations of the course provided to the National Fish and Wildlife Foundation indicated that the students were enthusiastic about continuing the practicum concept. The course promoted the idea of practice-based research involving project designers and construction practitioners as well as outside observers. Students noted that they could provide some objective observations of project performance but were ultimately constrained by their lack of familiarity with a design process and tools. The feedback we received was that they recommended more integration of theory and practice in university curricula.

A participating graduate student, Mark Spencer, and I wrote up a review of our experience for this practicum that was published in the summer 2000 issue of *Ecological Restoration* (Jordan 2000; Riley and Spencer 2000). Our conclusion was that the lack of positive reception from professors was not going to allow this concept to sustain itself. Concerns expressed by academics included the creation of competing courses with the regularly scheduled courses. Concerns also included the distrust and lack of control over what the campus professors thought may be taught, a distrust of active versus passive restoration, and the discomfort with nonprofit organizations that integrate a variety of participants with ranging educational and skill levels from high school graduate equivalency degrees to advanced degrees. The dream that Bill Jordan and I had of integrating different levels of education, field skills, construction knowledge, and experience into the educational process was not to be in these circumstances. The experiment started at Village Creek ended at Village Creek.

Beaver Restoration Crews on Alhambra Creek, Downtown Martinez, 2008

Location: Escobar and Castro Streets, Martinez, California
Project length: Approximately 1,000-foot area

Project History

On her walk through downtown in 2006, Heidi Perryman learned from a stranger that beaver had been sighted in Alhambra Creek, which flows through the center of Martinez (fig 3.68). Perryman, a native of Martinez, walked down Escobar Street, two blocks from Main Street, and stopped where the road crosses the creek to see for herself. There it was, a beaver dam in the business district. Heidi began to write articles for the local *Martinez News Gazette* to inform more residents about the family of beavers with a mother, father, and four kits who had moved into the mouth of the creek from the nearby Carquinez Strait connected to the Sacramento–San Joaquin delta. By the summer of 2007, business was booming at the nearby coffee shop on Main Street where locals went to watch the beaver. By this time, the Martinez City Council had become aware of the beaver, and in a summer meeting, council member Mark Ross advised leaving the beaver alone and "letting nature takes its course."

An ominous sign, however, appeared in September when the city council directed the city engineer to hire an engineering consulting firm to assess whether the beaver dam would create a flooding hazard. The report's conclusions that there could be a flood impact caused by the dam ultimately led to a populist beaver protection movement in Martinez. The civil engineering firm hired to assess the impacts of the beaver dam used an HEC-RAS model that included upstream and downstream bridge crossings to establish floodwater surface elevations with and without the beaver. The model indicated that the creek channel has the capacity to contain the 1-in-10-year food (2,400 cfs) and that the dam elevated the flood levels by 2 feet for this flood event, thereby creating overbank flows near Escobar Street (Tucker 2007). The city officials reacted to the report by declaring the beaver a public health and safety hazard and recommending extermination. (This reaction, in which officials frequently consider beaver a pest species, was common throughout the United States.) Council member Ross negotiated a compromise to work with the California Department of Fish and Wildlife to avoid immediate extermination of the beaver and instead procure a permit to relocate two of the beaver. When the news of a potential beaver relocation reached the townspeople, citizens gathered in a candlelight vigil held at the beaver dam. A November 7,

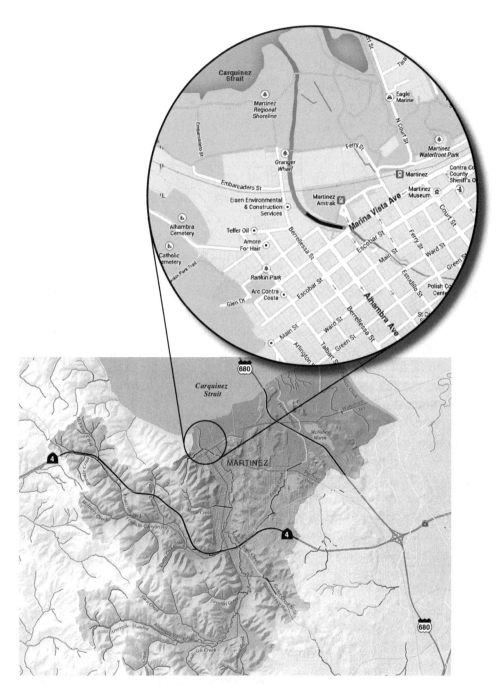

Figure 3.68 Alhambra Creek flows through the city of Martinez in Contra Costa County into wetlands at the edge of Carquinez Strait. *Credit: Lisa Kreishok.*

2007, city council meeting was designated as the forum at which decisions were to be made about how to address the beaver issue (Perryman 2014).

When that meeting convened, two hundred people or more crammed the city council chambers. The predictable result was that the council formed a committee to study the issue and then directed it to provide a recommendation to the city. The message that the public sent to the city and its committee was that they needed to find a way to manage the beaver with nonlethal means. With the forming of the City of Martinez beaver subcommittee, Heidi also formed the group Worth a Dam to organize the pro beaver citizenry (Perryman 2014).

By early January 2008, Skip Lisle, a beaver management expert from Vermont, was flown into Martinez at the urging of beaver advocates to assess the situation and make timely actions to avoid potential winter flood damages. Lisle installed a Castor Master as shown in figure 3.69 designed to control the flooding potential of the beaver dam. The top portion of the dam was removed, and the device was installed to hold the elevation of the pond created by the dam at an acceptable level. The hydraulic model of these conditions indicated that the Castor Master would lower the water surface elevation behind the dam by 2 feet and neutralize the dam's impact on flooding elevations at the most frequent flows. At the twenty-five-year flood, the influence of the dam is overridden by channel dimension limitations and bridge backwaters (Lindley and Nilsen 2007; Perryman 2014).

The subcommittee's final report to the city was released in March 2008 before another packed audience of citizens. The city's beaver subcommittee split into fifteen study areas and produced a report covering hydrology and flood management, water quality, stream bank stabilization, potential impacts of beaver on existing wildlife populations, beaver population control and dispersal, projects costs and grant possibilities, among other topics. The hydrology and flood management section identified different options for management of the dam, including an emergency removal strategy, excavating a flood terrace between Escobar and Marina Vista bridges to compensate for flood capacity losses, installation of a bypass pipe to convey flood flows around the dam site, construction of a floodwall or berm along the west side of the creek and regrading of the creek topography to direct flood flows downstream into a public park toward the east, construction of a detention basin, and combinations of some of these options (City of Martinez, California 2008; Avalon 2014).

Seven of eight subcommittee members—who represented the county and city public works departments, creek-side property owners, downtown merchants, and Worth a Dam representatives—recommended keeping the beavers alive. With peace ready to break out, another development—in the form of a letter from an attorney for a property owner with a business on the bank of the creek near the dam—threatened the resolution of the beaver habitat. The property owner alleged

(a)

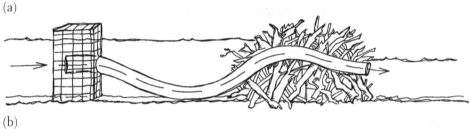

(b)

(c)

FIGURE 3.69 (a) The Castor Master. Photo credit: Skip Lisle, president of Beaver Deceiver International. (b) A sketch of the device. *Credit: Lisa Kreishok*, adapted from the Utah Division of Wildlife. (c) The Castor Master sends the stream flow silently through the top of the beaver dam.

that the beaver had tunneled under a retaining wall and were endangering the structural safety of his building (Langbehn 2008).

Another engineering firm hired by the city reported that the stream bank was eroding around the back of the retaining wall and that this stream bank therefore constituted a hazard to the building on top of the bank. The city declared the need for an emergency project and recommended installation of a metal sheet wall along the bank adjacent to the beaver dam. The emergency declaration circumvented standard environmental review and pubic noticing of the project under state law. The emergency status of the project was contested by Worth a Dam's attorney, but the court sided with the city. A stream consultant to Worth a Dam, Laurel Collins, assessed the conditions in the creek and did not find atypical creek conditions near the retaining wall and found a substantial below-grade concrete footing for the wall in question, indicating a stable, nonemergency situation. A city employee also found past photos of the site that indicated that sheet piling had been previously added to reinforce the wall. The city nonetheless acted with much engineering conservatism and placed another protective sheet wall at the site. The Worth a Dam citizens placed monitors at the site during construction to ensure that damage did not occur to the dam or lodges and that the beaver were not harassed. The beaver located future stream bank lodges away from this location.

From this point on, conflicts have not returned to the beaver site. High water washed a lodge out in 2011, but the beaver continued to thrive with a mother, father, and three kits. Since 2011, at least fifteen kits have dispersed from the site to new locations.

Project Design and Construction

Protecting and managing beaver in urban streams to meet public objectives to incorporate a wildlife feature to their downtown, enrich their children's experiences, and create a local identity fall squarely in the category of passive restoration. The need to reduce conflicts between the beaver and the habitat they create requires intervention to address flooding issues; therefore, this type of active intervention by humans is needed in most urban streams.

The Castor Master illustrated in figure 3.69 was constructed and fit into the unique situation created by the urban Alhambra Creek, which is tidally influenced and conveys fluvial flows on an intermittent basis. Bridge crossings control the channel hydraulics for flows at and above the twenty-five-year flood level. The in-channel Castor Master was able to lower the water surface elevation for the most frequent flows by lowering the effective height of the dam and the pool behind it. The Castor Master and another product known as the Beaver Deceiver are the creation of Lisle, who invented these devices in the mid-1990s to reduce conflicts between beaver dams and human environments. Wetland and river management and restoration communities promote and disseminate information on the use of these devices because of the importance of beavers to creating productive wetland and floodplain habitats. A diversity of organizations, agencies, and companies—such as the Fund for Animals, the Humane Society of the United States, state wetland associations, state and regional wildlife management agencies and companies like Lisle's—assist property owners, highway departments, and local officials reduce conflicts between beaver and humans (Simon 2006).

Lisle continues to evolve his designs for these two systems based on installation and performance experience. The products are part process and part hardware because they need to be correctly designed and fit to different environments to ensure success. The Beaver Deceiver is in use to prevent the occupation of beaver in culverts and other constructed outlets. It employs mostly a fencing system that allows beaver movement in the stream but prevents damming behavior inside the culverts. Later, piping was added to the Beaver Deceiver fencing to moderate flows through the beaver-occupied area. The Castor Master was invented to reduce the elevation of beaver dams situated in stream channels. This system of pipes with filters is designed to lower flooding elevations behind the dams. The Castor Master draws water from upstream into a pipe protected from clogging by a filter device, and the water cascades over the downstream dam at a selected height. By "caging" the intake with the filter and submerging the drainpipe so that it flows silently, beavers are unable to detect or prevent the leak of water that goes through their dam in the pipe. Beavers will build dams to meet the elevation of stream flows around their dam. The Castor Master regulates the height of dams by keeping the pool of the water behind the dams at a selected elevation, disguising that flows are leaking through the dam. This intervention discourages the beavers from building the dam higher to meet the higher pool elevations that would occur without this it (Utah Division of Wildlife Resources 2013; Lisle 2014).

The Alhambra Castor Master uses a round, 6- to 7-foot-diameter cylinder that functions as a filter to keep the flow pipe from clogging. The latest designs in filters (which look like metal "cages" with a bottom and a top) are square and reinforced with wood. The distances between the filter and the piping to the dam are custom fit to each stream situation. Typical prices are $2,500 for Beaver Deceivers and $1,000 to $2,500 for Castor Masters. Prices and designs vary by numbers of pipes used and amount of labor involved (Lisle 2014).

Alhambra Creek experienced high flows in 2011 that dislodged the filter. It was retrieved and reinstalled. The management objective in this urban stream is to regulate low to moderate stream flows. Floods in confined urban streams will wash out the beaver dams. Once natural processes remove the "roughness" elements and constrictions caused by sticks and branches, the dam is more or less self-regulating based on the ability to break apart and flow downstream in flood flows. The beavers rebuild after the flood, and the cycle goes on. Martinez has strategically addressed flooding in its business district by removing hydraulic constrictions at the mouth of the creek and working with the railroad to elevate the trestle span over the creek at the railroad station. Marsh plains at the creek's mouth were extended, and stream floodplains and bridge crossings have been widened in the lower section of creek. These modifications make it less likely for beaver dam material to discharge downstream in a flood and create debris jams. Even though the risks of debris racking are never removed, they can be reduced to acceptable levels. That is also the management strategy being applied in the agricultural wine-growing areas of the nearby Napa River and in Napa Creek in downtown City of Napa in which the perception of conflict between beavers and landowners is mitigated by knowledge that flood flows in this relatively constrained floodplain system will wash dams out and thereby reduce property damage risks during flood flows (Sarrow 2013). Ironically, the flashiness of urban stream systems assures this type of cycle in which dams are washed out during floods and return during the low-flow conditions. Figure 3.70 shows a beaver rebuilding a dam in Alhambra Creek. The usefulness of the investment in the Castor Master and Beaver De-

FIGURE 3.70 An Alhambra Creek beaver constructs a new dam. *Photo credit: Worth a Dam.*

ceivers for moderating the most common flows can complete this management scheme.

Project Lessons and Significance

The significance of this project is its use as a prototype for other cities and property owners for a management scheme that allows beavers to recolonize a region where they once influenced the ecological diversity. Another significant lesson from the Alhambra Creek case is that wildlife occupied the creek environment as a result of the new habitat created by the beavers, which had not been recorded previously. The basis of comparison of before and after beaver wildlife occupation used an environmental impact report inventory of aquatic species in lower Alhambra Creek that was prepared in 2005, before the beavers occupied the stream in 2006. After beaver occupation, wildlife and fish sightings were recorded with photos that include species not noted in the 2005 report investigation. These species included threatened species or species of concern to the state. The observations include steelhead, lesser scaup, western pond turtle, tule perch, Sacramento spittail, mink, river otter, and muskrat. New species of birds include hooded merganser, common merganser, black-crowned night heron, American bitten, western grebe, Clarke's grebe, pied-billed grebe, western grebe, common snipe, American coot, and great blue heron. This tidal environment does not typically have a fluvial contribution to flows in the six-month dry season, and the beaver dams and lodges have created a perennially flooded area that enables the location to support a greater biological diversity (Perryman 2014). This phenomenon is corroborated by researchers in wildlife biology and water quality who describe the ecological and water quality values associated with beaver occupying streams (Pollock, Press, and Beechie 2004; Lazar et al. 2015).

As a result of the Alhambra Creek experience, Heidi Perryman became instrumental in gathering a group of western scientists to address the issue of whether beavers historically occupied San Francisco Bay Area watersheds. The newly formed California Beaver Working Group began to address this issue in 2012. The seven-author paper that resulted uses a diversity of sources to conclude that beavers have been ecological forces before and during early Spanish and European settlements in the San Francisco Bay Area and other Northern California locations. This research paper reversed the misinformation provided by older, less rigorous studies that concluded that beavers were not native to California. This study has significant consequences because the State of California needs such information to reclassify the beaver from a nonnative pest species to a native species that should have a management plan developed for beaver-based ecological recovery as well as management for human considerations. This study has helped raise awareness of beaver as a key species for ecological recovery for professionals involved in stream, floodplain, and fisheries restoration (Lanman et al. 2013). Worth a Dam continues to educate communities in California and interact with state wildlife officials to prevent the routine issuance of depredation permits to shoot or remove beavers. A recent report by three federal agencies and a university now provides a more up-to-date handbook to guide managing beavers as environmental assets in our urban and rural environments (Pollock et al. 2015).

The Martinez beaver case includes the remarkable public response to add wildlife to the urban living experience, despite inconveniences or threats to public safety. The city promotes its new image as the beaver city, and the city council acknowledges that the beavers are a draw for regional visitors to the downtown business district. An annual beaver festival sponsored by Worth a Dam since 2008 attracts approximately five thousand people from five counties to downtown Martinez each year. Beaver ecology and biology has provided experiential science education for local schoolchildren and community college students. The beaver festival has put a sleepy town on the map for visitors and tourists. Council member Ross asks, "Where else can you see beavers within 15 yards of parking meters?"

This case teaches us that we can enjoy the new ecosystems that beavers can bring to urban environments and inexpensively prevent conflicts with human needs such as the prevention of flooding. Once flow devices such as the one used by Martinez are installed, they typically require only minor periodic maintenance. The advantage is that the use of such devices enables a sustainable beaver population, which will keep other dispersing beavers away and will thereby avoid an expensive, ongoing beaver trapping program, an advantage that lethal extermination projects do not have (Utah Division of Wildlife Resources 2013).

Perryman advises that if a similar conflict develops between urban dwellers and beavers or even other wildlife species, the best defensive strategy for wildlife advocates is to conduct as broad an education effort and reach as many people as possible with factual information about the real versus perceived risks. An educated public is the most potent tool to protect the urban wildlife. Wildlife advocates need to take their role seriously in being open to potential human-wildlife conflicts and assist in finding the solutions to them (Perryman 2014).

Other Alhambra Creek Watershed Projects

Two other projects on Alhambra Creek have addressed flood reduction, bank stabilization, and ecological restoration. The first project entailed the elevation of the railroad over the mouth of the creek and expansion of the wetlands and floodplain as the creek flows past the railroad station to the Carquinez Strait. The other project involved school students who turned a stream bank restoration project into a focus of the school curriculum.

RAILROAD ELEVATION AND WETLAND RESTORATION
Location: Martinez Railroad Station
Project length: 2,000 linear feet

Alhambra Creek serves as an example of the practice of good engineering to reduce flood damages while protecting and establishing new creek habitat. The railroad tracks located at the mouth of the creek as it enters San Pablo Bay were responsible for a significant portion of the flooding problems upstream in the town of Martinez. When the flood flows hit the low-placed Southern Pacific Railroad track soffit, the water backed up into downtown. County public works official Milt Kubick was not famous for his environmental consciousness, but he understood that flooding in Martinez could not be lessened without addressing that the railroad tracks were too low over the creek and the bridge span too narrow. Modifying the upstream creek channel would have no purpose if this hydraulic constriction was not removed at the end of the creek system. The railroad was convinced that chronic flooding of the train station at this location was not to its advantage, and Milt, using some leverage that he had in his county position, convinced the railroad to add 100 feet more to a 50-foot railroad bridge span over the creek and raise the soffit of the bottom of the bridge. The Contra Costa County Flood Control District worked with the railroad in 1998 to elevate the tracks and restore both the tidal prism and wetlands as well as the fluvial flows coming from up stream. A salt marsh enhancement project directly downstream at Grainger's Wharf included widening the creek flood and marsh plains, providing a secondary outlet for flood flows to Carquinez Strait. This enhancement project was complimented with the downtown improvement project, which widened the creek corridor, created floodplain benches, and widened a number of bridge crossings between Mira Vista and Green Streets. The Alhambra Creek watershed council, which originally formed in the 1980s and was reactivated after the flood damages of 1997, coordinates county, city government, and citizens who have developed a management plan for the watershed. This very comprehensive plan was finished in 2001 and combines flood damage reduction strategies, ecological restoration, and protection and emphasizes the removal of hydraulic constrictions and protection of the creek ecosystem as an amenity to the town (Tucker 2007, 2008; Avalon 2014).

The Martinez Adult School Stream Bank Restoration
Location: 600 F Street, Martinez, California
Project length: 500 linear feet

Upstream on Alhambra Creek is a pilot project supported by the county flood control district that addresses another common urban creek management need: the stabilizing of failing creek banks. The Martinez Adult School located on the banks of Alhambra Creek featured a channel with failed gabions and wood retaining walls. The ruins of the gabions were strewn along the channel, and property owners on the opposite bank were seriously considering suing the school district for damages. With the county's support, the Urban Creeks Council worked with the school district and used this pilot project to help illustrate better creek bank stabilization methods using soil bioengineering. The 2005–2006 project used only brush layering and willow post installation, without rock or other hard fortifications. The project involved regrading the failed banks to 2-to-1 side slopes. The January 2007 flood, which was estimated by the county to be a 1-in-40-year discharge for this watershed, did not destabilize the newly installed soil bioengineering project, which helped establish the site as a project to emulate. The ancillary benefit was the educational opportunity for students attending the Environmental Studies Academy located at the adult school. The environmental curriculum was developed to create a service-learning experience for students who were not successful in achieving at the local high school. The Urban Creeks Council involved the teachers and youth in implementing the project, and the newly motivated students won a national education award (Vukman 2013).

References

Strawberry Creek

Brand, William. 1989. "Battle for the Brooks." *Oakland Tribune*, January 12, A-1, A-3.

Charbonneau, Robert. 1987. "Strawberry Creek Management Plan." Office of Environmental Health and Safety, University of California, Berkeley.

Charbonneau, Robert, Stephanie Kaza, and Vincent Resh. 1990. "Strawberry Creek: A Walking Tour of Campus Natural History." Office of Environmental Health and Safety, Botanical Garden, and Department of Environmental Science, Policy and Management, University of California, Berkeley.

City of Berkeley, Department of Public Works and Design Section. 1982. "Construction and Planting Plans." *Strawberry Creek Park Project*, file 406.

Edlund, Lee. 1988. "Strawberry Creek Goes Biotech." *Daily Californian*, October 18, 7.

Mason, Gary. 1993. Project Designer, Wolfe Mason Associates. Interview. Berkeley, CA.

Montgomery, William. 1993. Director, City of Berkeley Parks, Recreation and Waterfront Department. Interview.

Owens Viani, Lisa. 1999. "Restoring Urban Streams Offers Social, Environmental, and Economic Benefits." In *Sustainable Use of Water, California Success Stories*, 283–303. Oakland, CA: Pacific Institute for Studies in Development, Environment and Security.

Pollock, Sarah. 1989. "Bringing the Waters Back to Life." *Urban Creek Restoration* 12 (6): 9–13.

Powell, Kevin. 1991. "The Free the Creek Movement." *Landscape Architecture* 81 (1):47–48.

Schemmerling, Carole. 2013. Commissioner, City of Berkeley Parks, Recreation and Waterfront Department; cofounder, Urban Creeks Council. Interview. Berkeley, CA.

Wolfe, Doug. 1987. "Recreating a 'Natural' Riparian Environment, or Getting the Creek Out of the Culvert." *Proceedings of the Second Native Plant Revegetation Symposium*. San Diego, CA, April 15–18.

Glen Echo Creek

Alameda County Flood Control and Conservation District. "Plans for Replacement of Slope Protection on Glen Echo Creek, Zone No. 12 Project." File CB-657.

Estes, Leslie. 2013. Program manager, Watersheds and Storm Drainage, City of Oakland Engineering Department. Interview, May.

Feng, Arleen. 2013. Volunteer steward, Glen Echo Creek; member, Friends of Glen Echo Creek, a project of the Piedmont Avenue Neighborhood Improvement League. Interview, June.

Hardin, Garrett. 1968. "The Tragedy of The Commons." *Science* 162 (3859):1243–1248.

Newhall, Barbara Falconer. 1986. "Oakland Rediscovers Its Webs of Water." *Oakland Tribune*, April 13, Lifestyle, section B, 1, 4.

Pollock, Sarah. 1989. "Bringing the Waters Back to Life." *Urban Creek Restoration* 12 (6): 9–13.

Starr, Kevin. 1989. "Oakland, Rebirth of The Sylvan Suburb." *Image*, March 12.

Wilson, Marlene. 1992. Founder, Friends of Glen Echo Creek; member, Oakland Heritage Alliance; volunteer, Piedmont Avenue Improvement League. Interview.

Winemiller, Valerie. 2013. Member, Steering Committee of the Piedmont Avenue Improvement League. Interview, June.

Wolin, Fred. 1993. Associate engineer scientist. Interview. Oakland, CA.

Blackberry Creek

Akagi, Danny. 2013. Engineer, City of Berkeley, California. Personal communication to author.

Alameda County Public Works Department. 1995–2005. Daily Rainfall, gauge 95d-UCBG. Berkeley, CA.

Askew, Mimi W. 1996. "Review of Blackberry Creek Daylighting Project." Project for Landscape Architecture Department, University of California, Berkeley. Water Resources Center Archives.

Berkeley Architectural Heritage Association. 1996. "What's New with Thousand Oaks School." 4 pgs. Berkeley, CA.

————. 2000. "Berkeley Architectural Association Plaque Project." Berkeley, CA.

Dunne, Thomas, and Luna Leopold. 1978. *Water in Environmental Planning*. San Francisco: Freeman.

Gerson, Stephanie, Jane Wardani, and Shiva Niazi. 2005. "Blackberry Creek Daylighting Project, Berkeley: Ten-Year Post Project Appraisal." Restoration of Rivers and Streams Series. Landscape Architecture Department, University of California, Berkeley. Water Resources Center Archives.

Gray, Donald, and Andrew T. Leiser. 1982. *Biotechnical Slope Protection and Erosion Control*. New York: Van Nostrand Reinhold.

Imanishi, Junichi. 2000. "Evaluation of the Blackberry Creek Daylighting Project." Project for Landscape Architecture Department, University of California, Berkeley. Water Resources Center Archives.

Leopold, L. B. 1994. *A View of the River*. Cambridge, MA: Harvard University Press.

Leopold, Luna, M. Gordon Wolman, and John Miller. 1964. *Fluvial Processes in Geomorphology*. San Francisco: Freeman.

Rantz, S. E. 1971. "Suggested Criteria for Hydrologic Design of Storm-Drainage Facilities in the San Francisco Bay Region, California." *San Francisco Bay Region Environment and Resources Planning Study*, Technical Report No. 3. Prepared in cooperation with the US Department of Housing and Urban Development Office of the Assistant Secretary for Research and Technology. Menlo Park, CA: US Department of the Interior, Water Resources Division, Geologic Survey.

Riley, A. L. 1994. "Restoration Design for Blackberry Creek at Thousand Oaks School." Design Memorandum for Urban Creeks Council, Berkeley, CA.

————. 1995. "Construction Notes and Surveys." In files of Southwest Coalition to Restore Urban Waters. Waterways Restoration Institute. Berkeley, CA.

Riley, A. L. 2013. *Restoring Streams in Cities: A Guide for Planners, Policymakers, and Citizens*. Washington, DC: Island Press.

Riley, A. L., and Drew Goetting. 1999. "Blackberry Creek Survey." Berkeley, CA: Waterways Restoration Institute.

Schemmerling, Carole. 1984. "Urban Creek Restoration Bill Is Law." *Sierra Club Yodeler*, November, 7.

————. 2013. Commissioner, City of Berkeley Parks, Recreation and Waterfront Department; cofounder, Urban Creeks Council. Interview. Berkeley, CA.

Szumski, Dan. 1995. *Blackberry Creek Restoration, Thousand Oaks School*. Isleton, CA: Dan Szumski and Associates.

Thousand Oaks Park Society. 1994. "Creating a New Park in Berkeley." 1-pg. pamphlet. Berkeley, CA.

Thousand Oaks School. 1994. "Restoring Blackberry Creek at Thousand Oaks School." 2-pg. pamphlet. Berkeley, CA.

United Nations International Trade Center, Coir Board of India, Robbin B. Sotir and Associates. 1991. "1991 North American Coir Geotextile Seminars." Prepared for United States Coir/Geotextile Conference, San Francisco.

Vigil, Delfin. 2003. "What's Not Working." ChronicleWatch: Working for a Better Bay Area. *San Francisco Chronicle*, October 26.

Baxter Creek

Batchelder, Steve. 2011. "Review of Poinsett Park Trees and Plant Materials." City of El Cerrito, CA, Public Works Department.

Brady Landscape Architecture. 1997. "Plans for Baxter Creek." Prepared for the City of El Cerrito, CA.

Butt, Tom. 2015. "Richmond Wins $5.1 Million Grant for Affordable Housing and Baxter Creek Restoration." Tom Butt E-Forum.

Charbonneau, Robert, and Vincent Resh. 1992. "Strawberry Creek on the University of California Campus: A Case History of Urban Stream Restoration." *Aquatic Conservation Marine and Freshwater Ecosystems* 2:293–307.

Chin, Anne, Shannah Anderson, Andrew Collison, Barbara Ellis-Suga, Jeffrey Haltiner, John Hogervorst, G. Mathias Kondolf, Linda O'Hirok, Alison Purcell, Ann Riley, and Ellen Wohl. 2009. "Linking Theory and Practice for Restoration of Step Pool Streams." *Environmental Management* 43:645–661.

Chin, Anne, Alison Purcell, Jennifer Quan, and Vincent Resh. 2009. "Assessing Geomorphological and Ecological Responses in Restored Step-Pool Systems." Geological Society of America Special Paper 451:199–214.

City of El Cerrito. 1994a. "Storm Drain and Creek Restoration Opportunity Sites: Report No 3." Prepared by Brady and Associates and Harris and Associates.

———. 1994b. "Storm Drain and Creek Restoration Program: Construction Activity Storm Water Pollution Prevention Plan, Report No 7." Prepared by Woodward Clyde Consultants and Harris and Associates.

Friends of Baxter Creek. 1998a. "Help to Save the Creek." Letter to Friends of Baxter Creek, El Cerrito, CA.

———. 1998b. Letter to Mayor Rosemary Corbin, City of Richmond, CA.

———. 1998c. Memorandum to the El Cerrito Redevelopment Advisory Committee, El Cerrito, CA.

Goetting, Drew. 2013. Principal, Restoration Design Group. Personal communication to author. Berkeley, CA.

Hruby, Thomas. 1999. "Assessment of Wetland Functions: What They Are and What They Are Not." *Environment Management* 23 (1):75–85.

Joint Watershed Goals Statement. 1995. Signed by representatives of the cities of Richmond, El Cerrito, Albany, Berkeley, and East Bay Regional Park District.

Jones, Carolyn. 2011. "Richmond Nurseries Make Way for Housing." *San Francisco Chronicle*, March 30, C3.

Kusler, Jon. 2004. "Assessing Functions and Values." In *Report I: Wetland Assessments for Regulatory Purpose*. Prepared by the Association of State Wetland Managers Institute for Wetland Science and Public Policy.

La Force, Norman. 2013. Mayor, City of El Cerrito, 1995–1999. Personal communication to author.

Mintz, Melanie. 2014. Community Development Director, City of El Cerrito, CA. Personal communication to author.

Ortiz, Yvetteh. 2014. Director of Public Works, City of El Cerrito, CA. Written communication.

Owens Viani, Lisa. 1996. "Rehab, Daylighting Poinsett Creek." *Estuary* 5 (5).

———. 1997. "Daylighting a Creek." *Urban Ecologist* 1: 9.

———. 2000a. "A Cultural and Natural History of the Baxter Creek Watershed." Richmond, CA: Aquatic Outreach Institute.

———. 2000b. "Restoring an Urban Stream: Baxter Creek Case Study." San Francisco State University Research Project.

———. 2009. "Plants That Did and Did Not Survive at Booker T. Anderson Park Baxter Creek Project." Memorandum to the San Francisco Bay Regional Water Quality Control Board, Oakland, CA, November 23.

———. 2013. Founder, Friends of Baxter Creek. Personal communication to author.

Purcell, Alison. 2004. "A Long Term Post-Project Evaluation of an Urban Stream Restoration Project." University of California, Berkeley. Water Resources Center Archives.

Purcell, Alison H., Carla Friedrich, and Vincent Resh. 2002. "An Assessment of a Small Urban Stream Restoration Project in Northern California." *Restoration Ecology* 10 (4): 685–694.

Riley, A. L. 1998. *Restoring Streams in Cities: A Guide for Planners, Policymakers, and Citizens.* Washington, DC: Island Press.

Robbins, Jim. 2013. "The Year of the Monarch Didn't Appear." *New York Times*, November 24, 9.

San Francisco Bay Regional Water Quality Control Board. 2007. "A Rapid Assessment System for Riparian and Stream Corridors for California: A Discussion on How We Can Improve a Rapid Assessment for Streams." Oakland, CA: San Francisco Bay Regional Water Quality Control Board.

Scarpa, Lynne. 2014. Environmental manager, Stormwater Program, City of Richmond, CA. Personal communication to author.

Struve, Mori. 1998. "Comparison of Costs of Creek Restoration vs. Culvert Replacement on Site." El Cerrito Community Development Department, El Cerrito, CA.

Waterways Restoration Institute. 1996a. "Adjustments to Creek Restoration Project at Poinsett Park." Prepared for Harris and Associates, Berkeley, CA.

———. 1996b. "Grading and Bank Stabilization Plans for Poinsett and Ohlone Parks." Prepared for Harris and Associates, Berkeley, CA.

Village Creek

Asher, Melissa, and Kaumudi Atapattu. 2005. "Post-Project Appraisal of Village Creek Restoration, Albany, Ca. for Restoration of Rivers and Streams." Water Resources Center Archives, University of California, Berkeley.

City of Albany Watershed Management Plan. 1998. Technical Appendices. Albany, CA.

Covina, Gina. 2000. "The Angry Village." *East Bay Express* 22 (29):1, 8–11.

Isbill, Julie K. 1988. "University Village Creek Corridor Design." Professional report for the Landscape Architecture Department, University of California, Berkeley.

Jones, Sandis Humber. 1998. "Flood Study for Village Creek." Prepared for the City of Albany, CA.

Jordon, William R., III. 2000. "Barriers—or Bridges?" *Ecological Restoration* 18 (2):73.

Lee, Warren F., and Catherine T. Lee. 2000. *A Selective History of the University Village, The City of Albany, and Environs.* Albany, CA: Belvidere Delaware Railroad Company Enterprises.

Lichten, Keith. 2014. Permit staff, San Francisco Bay Regional Water Quality Control Board Section 401, 1999–2000. Personal communication to author.

Logsdon, Willis, and Connor McIntee. 2014. "Profile and Cross-Sections at Village Creek 2014." Watershed Stewardship Project of the California Conservation Corps, Sacramento, CA.

May, Loren. 2014. President of May and Associates, Sonoma, CA. Personal communication in the field at Village Creek.

O'Connor, Dennis. 2014. Principal of Habitat Concepts, Portland, OR. Personal communication to author.

Richards, Keith. 1982. *Rivers Form and Process in Alluvial Channels.* London and New York: Methuen.

Riley, Ann L., and Mark Spencer. 2000. "Teaching Environmental Restoration: A University-Nonprofit Collaboration." *Ecological Restoration* 18 (2):104–108.

San Francisco Bay Regional Water Quality Control Board. 2000. "Albany Village Project and Request for a Report on Waste Discharge on Village Creek, Albany Village, Alameda County." Oakland, CA.

University of California, Berkeley. 1997. *University Village and Albany/Northwest Berkeley Properties Master Plan.* Project no. 912800. Prepared by the Office of Planning, Design and Construction, Physical Environmental Planning, Berkeley, CA.

———. 2004. *University Master Plan.* University of California Berkeley Facilities Services Final Draft.

US Coast Survey. 1865. *U.S. Coast Survey San Francisco Bay.* California Plane Table Sheet XXV, A. D. Bache, Superintendent.

Alhambra Creek

Avalon, Mitch. 2014. Past deputy director, Contra Costa County Public Works Department; chair, Alhambra Creek Watershed Council; chair, City of Alhambra Beaver Subcommittee on Hydrology and Flood Management. Personal communication to author.

City of Martinez, California. 2008. "Beaver Subcommittee Report Final." March 3.

Langbehn, William L. 2008. "A Preliminary Geotechnical Evaluation of Alhambra Creek." Letter to Huguet Turner and Martinez Adams, February 18.

Lanman, Christopher, Kate Lundquist, Heidi Perryman, J. Eli Asarian, Brock Dollman, Richard Lanman, and Michael Pollock. 2013. "The Historical Range of Beaver (*Castor Canadensis*) in Coastal California: An Updated Review of the Evidence." *California Fish and Game* 99 (4): 193–221.

Lazar, Julia, Kelly Addy, Arthur Gold, Peter Groffman, Richard McKinney, and Dorothy Kellogg. 2015. "Beaver Ponds: Resurgent Nitrogen Sinks for Rural Watersheds in the Northeastern United States." *Journal of Environmental Quality* 44 (5): 1684–1693.

Lindley, Mark, and Christian Nilsen. 2007. "Martinez–Alhambra Creek Beaver Dam Assessment." Prepared for the City of Martinez Engineering Department, Philip Williams Associates, San Francisco.

Lisle, Skip. 2014. President of Beaver Deceiver, International, VT. Personal communication to author.

Perryman, Heidi. 2014. Founder of "Worth a Dam" Citizens Organization, Martinez, CA. Personal communication to author.

Pollock, M. M., G. Lewallen, K. Woodruff, C. E. Jordan, and J. M. Castro, eds. 2015. *The Beaver Restoration Guidebook: Working with Beaver to Restore Streams, Wetlands, and Floodplains.* Version 1.0. Portland, OR: US Fish and Wildlife Service.

Pollock, Michael, G. R. Press, and T. J. Beechie. 2004. "The Importance of Beaver Ponds to Coho Salmon Production in the Stillaguamish River Basin, Washington, USA." *North American Journal of Fisheries Management* 24: 749–760.

Sarrow, Jeremy. 2013. Watershed and flood control resources specialist, Napa Public Works Department, Napa, CA. Personal communication to author.

Simon, Laura. 2006. "Solving Beaver Flooding Problems through the Use of Water Flow Control Devices." In *Proceedings of the 22nd Vertebrate Pest Conference*, edited by R. M. Timm and J. M O'Brien, 174–180. Davis: University of California Press.

Tucker, Timothy. 2007. "Alhambra Creek Beaver Dam Assessment." Philip Williams Associates Memorandum Report. Prepared for the City of Martinez, CA.

———. 2008. "Alhambra Creek Beaver Dam Management Options." Memorandum report prepared for the City of Martinez, CA.

Utah Division of Wildlife Resources. 2013. "Beaver, Best Management Practices."

Vukman, Mike. 2013. Chair, Statewide Steering Committee, California Urban Streams Partnership; past restoration program manager, Urban Creeks Council, San Leandro, CA. Personal communication to author.

Chapter 4

What Neighborhood Projects Teach

What have we learned from these cases? Highly impacted urban environments can support dynamic, functioning stream systems that can support fish and wildlife habitat. Most degraded stream systems require an active restoration approach to return stream processes to re-create channels, floodplains, and riparian resources. The exciting relatively new field of historic ecology has increased our awareness of the ecosystems that used to exist and the functions they performed. Typically, we cannot re-create these ecosystems in developed urban areas, but we can create new environments that can emulate some of the past ecological processes and functions. Central to re-creating some of the functionality is advocating for adequate floodplain area so that the streams have room to adjust and re-form. Some of these re-created environments—such as meandering, single-thread channels through restricted floodplain corridors—can illicit derision from academia, which has the tendency to focus on the limitations of the urban landscape and the desirability of returning the historic landscape. From the perspective of needing to create alternatives to single-purpose flood and erosion control projects, however, the urban streams restoration movement has introduced viable environmental alternatives.

The lessons evolving from restoration design practice teach that we are generally at our best when we combine the different schools of restoration and not get bogged down in advocating one school at the exclusion of another. Hydraulic models—such as the commonly used HEC-RAS—can be practical tools to ensure that flood risk is being reduced rather than increased by a project proposal. The empirical tools—such as the use of analogs, hydraulic geometry relations, and regional curves—were useful to the primary design of restoration channel and floodplain dimensions. If regional curves are matched correctly with similar homogeneous geographies and microclimates, apply to environments that have adjusted to long-term urban conditions, and are not undergoing land use changes, they are very effective design tools for an urban setting.

Restoring riparian corridors needs to better focus on identifying appropriate reference plant communities to achieve a functioning riparian environment and desired survival rates. Planting projects can often best be carried out using a phased approach to increase success in creating a complex structure to the riparian corridor and species diversity.

It bears repeating that restoration projects are only as good as the available floodplain area set aside to support ecological functions. The two keys to reducing flood damages from urban streams are (1) the protection and acquisition of floodplain and (2) locating and removing hydraulic constrictions causing the overbank flows at culverts, trestles, and undersized bridges. Attempting to remake streams into straightened ditches for flood management purposes is a losing strategy.

Setting Project Objectives

An overriding objective for most of the neighborhood projects discussed in chapter 3 was to replace conventional stream engineering practices that channelized and hardened stream channels with practices that could support freely dynamic and unrestrained streams with sustainable native riparian corridors yet still address public safety needs. Counterintuitively to some, we wanted to show that by making streams more dynamic, they would actually be more stable and less prone to causing flood and erosion damages than their less-dynamic counterparts. Not until the Friends of Baxter Creek became involved with urban watershed restoration did the concept of restoring wildlife habitat at smaller reach-level urban projects become one of the primary purposes of the projects. If we compare the cases in this book with other neighborhood-scale projects from the larger San Francisco Bay Area, in which there were approximately forty watershed councils or friends of creek groups by 2003, we find that the cases are representative of the region-wide efforts. On a regional scale, most neighborhood-scale projects were formed primarily to address flooding and erosion issues and neighborhood environmental improvement, although a few were formed primarily to address recovering anadromous fish populations. The Friends of Baxter Creek saw a potential value in neighborhood-scale creek restoration efforts that had not initially occurred to us. One of the founders of this organization of citizens, Lisa Owens Viani, a bird enthusiast who was pursuing a master's degree in geography, set out to record bird populations at the newly completed Baxter Creek restoration on Poinsett Avenue. Owens Viani was inspired by this experience and, based on her incentive to create bird habitat in cities, proceeded to organize, fund, and implement the stream restoration project on lower Baxter Creek at Booker T. Anderson Park. We also became aware that bird habitat was a driving incentive for much of the stream, floodplain, and related wetlands restoration in the Portland, Oregon, area through the efforts of Mike Houck of Portland Audubon, who sponsored a series of "Country in the City" symposiums in the 1990s.

The ultimate change of awareness affecting urban stream restoration objectives came when it was demonstrated to us that aquatic mammals could return to urban centers in small watersheds. The appearance of beavers in Alhambra Creek in the City of Martinez's commercial district opened our minds to biological objectives that we had never before conceived. Reports of urban stream otters are now becoming more frequent in the San Francisco Bay Area. By 2013, the River Otter Ecology Project had conducted the first formal research of otters in the Bay Area. Using photos and videos taken over a two-year period in very urbanized settings, they confirmed six hundred sightings in areas that included Alhambra Creek in Martinez, Temescal Creek and Lake and Lake Merritt in Oakland, Wildcat Creek and Jewel Lake in Berkeley, and the Richmond Marina and Sutro Baths in San Francisco (Dearen 2013). By 2013, beavers were found in the main stem of the Napa River, and by 2015, they were residing in Napa Creek in the City of Napa downtown. Beaver are now reported to live in the City of San Jose in the Guadalupe River watershed. Reports of beaver occupying major urban rivers are increasing and count among them the occupation of the Anacostia River in Washington, D.C. (Griffin 2015)

It was our initial hope, as opposed to an actual objective, that reach-level projects could improve water quality. Post-project monitoring found that some important parameters such as dissolved oxygen and temperature were influenced by reach-scale projects but that improvement in benthic insect pollution-sensitive species and diversity was not occurring. As resources agencies began to formally recognize the role that healthy riparian corridors could contribute to water quality, water quality evolved into a more formal objective.

These reach-scale restoration projects were part of a chicken-and-egg issue when it came to setting restoration objectives. The Waterways Restoration Institute accidentally discovered steelhead under the Sixth Street culvert on Codornices Creek in 1996 while monitoring a recently completed reach-level project. This sighting was confirmed by another citizens' group, the Friends of Five Creeks, which was involved in invasive plant removal and landscaping in an upstream location a few years later. The Urban Creeks Council discovered steelhead when members were engaged in small stream projects for property owners on Alhambra Creek in the City of Martinez. Restoration objectives were discovered as a result of the restorationists' activities on the creeks.

Experience with neighborhood-scale projects helped create the expertise and momentum needed to tackle larger, regional-scale projects that approached a mile or more in length. Regional-scale projects on Wildcat Creek, Napa River, Napa Creek, and Codornices Creek represented innovations in large-scale, multi-objective flood risk reduction projects. The Napa River and Napa Creek projects are importantly distinguished from the Wildcat and Codornices Creek projects in that they represented an advancement from integrating environmental functions into a flood control trapezoidal right-of-way. Restoration objectives for the Napa River were itemized in the adopted Living River principles written by a team of scientists who used open discussions in forums, reflecting a significant public consensus to formally plan for environmental objectives. Flood risk reduction was attained more through restoration of floodplains and wetlands than by adding environmental functions to a flood reduction channel. The Napa River and Creek Living River principles covered everything from achieving fish habitat restoration to water quality improvements to returning geomorphic functions to the river and creek. In contrast, the Wildcat Creek project objective focused first on flood risk reduction because of overwhelming public consensus for this need to save a low-income community from chronic flood damages but pioneered the multi-objective principle of also delivering environmental protection and ecological restoration while reducing maintenance needs. This multi-objective approach was largely forced by the federal Endangered Species Act and organized public advocacy for more balanced project objectives. The Codornices Creek project objectives described in its planning documents were primarily focused on providing a flood control project for a new housing development, and it used an approach that the planners hoped would provide a more stable and sustainable as well as ecologically functioning channel, again within a trapezoidal right-of-way. The Village Creek project described here is another example of adding ecological function to what was originally conceived as a flood project channel.

How Projects Happen

Each neighborhood-scale project description begins with a story of how a citizen or groups of citizens organized to realize a restoration project. This trend characterized these projects in the region until the early 2000s, when East Bay cities and flood districts became familiar with availability of government restoration grant programs, recognized the ability of reach-scale projects to address flooding and erosion problems, and requested help from consultants or nonprofit organizations in preparing grant applications. The necessary first ingredients to realize restoration projects were the pioneers who took risks to attempt something new. The pioneers developed support by delivering better solutions to flood and erosion problems for local officials, and in many cases, they focused on the socioeconomic benefits to neighborhoods. The second crucial element to integrating restoration as a practice to be adopted by local governments was the pilot project approach. The first projects become the pilot projects that establish that the "restoration paradigm" is possible, that these types of projects are safe, and that the projects generate great community support and outside funding.

Starting in the 1980s, the Waterways Restoration Institute and Urban Creeks Council sponsored regular tours to these pilot projects and conducted workshops to describe how they were designed and funded. The tours and workshops increasingly involved engineers and planners from water districts, cities, and counties who were interested in solving common problems and conflicts involved with urban streams for which there was community support. The tours helped set up working relationships among the restoration community and the local governments and also served to help build a new network of organizations and agencies interested in pursuing stream restoration. In addition, the tours and workshops became an ongoing source of revenue important to sustaining the creek organizations. These pilot projects did not happen without support from the local jurisdiction where they occurred. The other critical component shared among these cases is identifying an influential person in a responsible agency within a public works or planning department, a commissioner, or a city or county council member who supports the implementation of the project.

A related central strategy was for creek restoration advocates to hold out the carrot of funding for the projects. One of the first pilot projects was a simple native planting on Glen Echo Creek, which was made possible through a federal water resources program administered by the state. The creek community realized that persuading local governments and water districts to budget local funds for a different type of project would be extremely challenging, which led to the statewide citizens' organizing effort to pass a bill to create a new state grants program. The urban streams legislation was written to appeal to the problem-solving needs for local flood and erosion damage reduction and required that the logical local agency and logical local citizens' organization be coapplicants to any grant. The success of the collaborative local pilot projects fueled a greater demand for stream restoration. Other state agencies began to financially support these projects, and over the next three decades, more funding sources developed from local, state, and federal sources. All the cases featured in this book except Alhambra Creek were enabled by grants. Some of the tours mentioned above were arranged specifically to inform local politicians of the benefits of the projects and to build relationships with state legislators who could commit to working to secure funding for the state grant programs we were using.

The other necessary ingredient was technical expertise. Unlike other branches of the environmental movement that historically blended ecological science and public policy, this new field of restoration required expertise and experience with on-the-ground application of civil engineering, geomorphology, hydrology, plant ecology, and horticulture. Again, a new generation of pioneers who could apply science to practice were needed. Critical to our movement in the San Francisco Bay Area was a new consulting firm, Phillip Williams Associates Ltd. Phil Williams is a fully credentialed engineer and scientist who was willing to buck conventional engineering practices, at times to the detriment of his business, and help define a new practice of restoration. Some of the new creek organization staffs also developed their technical expertise and were largely responsible for developing an awareness of soil bioengineering technology and practicing applied geomorphology. Two of the principals at Circuit Riders, a nonprofit organization, eventually formed a for-profit company that combined its early roots of organizing communities, raising grant funds, and developing community-based projects into a very successful business as Prunuske-Chatham, Inc. By the early 2000s, the field of stream restoration had become a business with good money to be made, and numerous consulting firms incorporated restoration work into their business model. Once a practice becomes a business, it signals the ultimate adoption of a concept.

Finally, restoration tends to be labor-intensive work, particularly the revegetation work. It is not a coincidence that the early origins of the urban stream restoration movement overlapped with the reemergence of the civilian conservation corps movement. The California Conservation Corps and the East Bay Conservation Corps (one of the many local corps that formed in the 1980s and 1990s) provided the capacity to hire crews of eight to ten workers who, after being trained in soil bioengineering practices, could collect, prepare, and install these systems. As restoration developed into more of a business, this aspect of restoration work became subsumed by contractors. Often, the contractors are viewed as more efficient and more accountable for construction standards than youth or conservation corps workers, and new prevailing wage jobs have been established. The benefits that have been lost from the model of using youth or conservation corps workers are entry-level jobs and training for at-risk youth and the hiring of youth from communities near the project locations as well as the loss of general support for the conservation corps organizations.

Restoration Design Methods Evolve

Chapter 3 walked us through the history of early projects in urban stream restoration, and the first impression we have of these early projects is that they are small reaches averaging 250 to 350 feet long. When starting something new, it is advantageous to begin small and build the confidence of the community, the project sponsors, and the project design and construction teams. By 2000, the East Bay practitioners had graduated to projects that were 900 feet long (at Baxter Creek at Booker T. Anderson Park), which is longer than the length of an average city block; 3,000 feet long (project on Codornices Creek); and 5,000 feet long (project on Wildcat Creek). The projects described in this book represent a sample of East Bay projects, including those on Seminary, Sausal, and Courtland Creeks in Oakland and early reach-scale projects on Codornices Creek in Berkeley. These projects all added to developing experience over time at the reach level. The reach-level restoration cases progress from the use of gabions and concrete slabs, replaced by the practice of soil bioengineering, erosion-control fabrics, and the construction of dynamic channels. Most important is that a design methodology emerged based on local information on prevailing stream processes and hydraulic geometry. Developing better designs incorporating stable channel widths and depths, width-to-depth ratios, and more stable channel lengths meant that we could design for dynamic ecosystems and replace the practice of building rigid, engineered channels. Hydraulic modeling, which had been the primary tool for designing urban channels, was either decided to be irrelevant or became the last step in design process, not the first.

Gabions and new products such as articulated concrete blocks and plastic geogrid systems were created and marketed by new businesses looking to satisfy engineers' evolving demand for replacements for concrete channels. Restoration practitioners, initially working at small scales, learned that rock and concrete and these concrete substitutes can be avoided in the effort to re-create dynamic channels and habitat if the designs create equilibrium "active" or bankfull channel dimensions, re-create channel lengths in balance with the valley slope and sinuosity, and add some floodplain space. With proper spacing and heights of the in-stream structures, steeper step pool channel types can be constructed to be both stable and dynamic. Soil bioengineering systems can then be used to complement the defense against the shear stresses on stream channels associated with high flood flows. These so-called soft bank stabilization methods are performing better after twenty to thirty years of observation than the riprap rock slopes or retaining walls that self-destruct because of channel incision, overnarrowing of the channel cross section, or other failure mechanisms. How can it be possible that this "natural" approach performs better in a built-up urban area within confined spaces?

While we learned that the historic channel types and planforms cannot be re-created, we also learned to apply the principles of hydraulic geometry to create new creek channels and floodplains, which balanced discharges, sediment supply, channel slopes, and shapes. The monitoring records of the projects in chapter 3 indicate insignificant changes in planform, cross sections, and slopes, and none of the projects experienced excessive erosion or deposition.

Applying the Schools of Restoration

Which schools of restoration discussed in chapter 2 were applied to these cases to achieve restoration design? Chapter 2 discussed the tensions that exist among differing perspectives on the scientific and engineering traditions that can guide restoration design. One school, the passive school, prefers managing watershed-scale influences such as sediment supplies, runoff rates, and initiating reforestation practices rather than addressing localized stream instabilities through projects that change stream conditions such as channel and floodplain modifications. This school contains the proponents of focusing on stormwater management first, before embarking on physical channel changes. The use of descriptions of stream processes—and a process-focused restoration approach, another related school—generally has not produced great detractors and has broad support. There are tensions among those who find it difficult to assign more value to quantitative evaluations of sediment loads and transport than to qualitative descriptions of sediment transport processes that they believe better capture complex watershed and sediment conditions.

Such tensions feed into the classic struggle between the use of analytical hydraulics models—the analytical school—to determine channel hydraulics and dimensions based on mathematical "governing equations" of continuity, flow resistance, and sediment transport versus the use of empirical field-based methods such as hydraulic geometry studies used to describe channel and floodplain dynamics and forms. The empirical data establish relationships among average channel dimensions and meander development and relationships between watershed drainage areas and channel dimensions, known as regional curves. The tensions within the biological sciences tend to be over the issue of whether we should be more concerned with population recovery or with biological and genetic diversity.

Passive School

There are two issues associated with the passive school in play with these cases. One is whether it would be best to approach urban stream restoration by engaging in stormwater management rather than restoration projects, and the other is to what degree we should emphasize self-recovery processes rather than construct channel and floodplain modifications.

The stormwater school proponents can take the position that stormwater run-off modifications should be accomplished before we undertake any reach-scale projects. They often use percent impervious watershed indices to predict habitat quality, and even restoration potential. Other researchers suggest that the portion of a catchment covered by stormwater piping is the best indicator of watershed degradation, and some suggest that the percent urbanization maybe a stronger correlation with biological decline. Yet others take issue with using these indexes to predict stream condition, habitat quality, and recovery potential and find that this assessment need is too complex to address using these indexes (Center for Watershed Protection 2003; Roesner and Bledsoe 2003; Walsh, Fletcher, and Ladson 2005; Chin 2006; Roy and Schuster 2009; Aparicio et al. 2011).

The cases presented in chapter 3 do not involve passive restoration, with the notable exception of the beaver-based restoration on Alhambra Creek in Martinez. A disastrous experience with a daylighting project on Codornices Creek used rough grading to remove a culvert, but the excavation was not based on creating an equilibrium design for a stable channel. The excavation was done with the hope that the stream would re-create an equilibrium form within the excavated space and ultimately provided a reality check on applying this form of passive restoration in an urban setting. As the channel slope was trying to adjust to the rough grading, the channel eroded down to an active gas line, and the banks caved in large sections, leading to emergency weekend meetings in the rain and flood flows to remedy the problems. This experience was lesson enough to use careful detailed design and advance planning in constrained urban sites, with the objective of finding a design channel geometry with the best chances of starting out as an equilibrium channel. The case studies described do apply process-based restoration strategies such as returning riparian corridors and woody debris, contributing materials for self-forming step-pools and aiding floodplain recovery. These strategies are distinguished from "passive" approaches in that they require construction and installation to achieve recovery of the processes and do not rely on larger-scale watershed modifications.

The Strawberry Creek project featured a before-its-time green stormwater design to replace the curb, gutter, and pipe approach to stormwater catchment. This design applied less expensive at-grade pervious drainage channels integrated into the park setting to catch and infiltrate site runoff. It avoided the typical stream channel destabilization caused by concentrated stormwater flows at pipe inlets to the creek, although it was executed at the scale that could only moderate the stream's hydrology locally.

The issue of abandoning stream restoration projects in favor of stormwater management was not a realistic alternative strategy for the project's proponents to achieve their objectives. The context for these projects is that even if aggressive, watershed-wide stormwater infiltration projects will be possible at some point, these good land management practices will not result in timely re-created equilibrium channels, re-created floodplains, restored meanders, step pools, or revegetated banks for streams that start out in culverts, or with failing riprap and retaining walls. Nor will they result in the timely recovery of creeks degraded by ditching.

What about the issue that percent imperviousness forecasts likely success or failure in achieving biological recovery or restoration objectives for active stream restoration projects? Estimates of watershed impermeability for the watersheds in which we were working range from 20 to 65 percent, which is certainly well beyond the 10 to 15 percent increase that the literature warns is the threshold for achieving a stable, ecologically functioning stream.

It is clear that the percent developed or impervious can be a good indicator for potential for biological recovery or lack of biological recovery, and the benthic assessment studies continue to conclude that reach-level restoration in cities is not changing the environment enough to get better results for diverse benthic insect assessments. What is typically missing in this body of research efforts, however, is information that can be used by a practitioner to understand the sources of the stressors so that restorationists and citizens' groups can address the sources of the problems. Certainly the new generation of green stormwater projects designed to slow, infiltrate, and treat runoff can provide multiple benefits and, presumably, if accomplished at the proper scale, can give us all a productive method to improve the biology of urban creeks. Urban stream improvement is more complicated than that, however.

Neighborhood-scale projects are one of our most powerful tools in developing public support for improving urban water quality. Four cases involve the public reporting of pollution and requests for public agency response: discharges from the hospital in the Glen Echo Creek watershed, reporting of sewage spills in Blackberry Creek, and chloramine discharges in Baxter Creek (frog kills) and Strawberry Creek (sewage pollution and fish kills). The reach-scale projects in the case studies have relatively high percent impermeability. That is an indicator of probable multiple urban stressors on the creeks, but does this indicator point us to the causes or solutions to environmental degradation? A review of stormwater literature indicates widespread agreement that we are only beginning to address causality and aquatic biotic degradation. Is the degradation mostly a function of changes in hydrograph such as volume and frequency of flows, changes in sediment, flashiness of flows, loss of vegetation cover, property owner channel modifications, or chemical and pesticide pollution? Is there a primary cause of degradation that we can generalize across many watersheds, or is each watershed a unique situation?

The Baxter Creek at Poinsett Avenue case introduced me to the widespread phenomenon of pesticide impacts on San Francisco Bay Area regional streams. My efforts to understand why the increase in benthic health scores after the restoration suddenly disappeared a few years later led me to the awareness of the large gap in information on the degree of chronic pesticide entry into urban streams. I also realized that we are not capable of approaching the issue of causality of urban aquatic degradation without closing this data gap. A number of Bay Area creeks are listed as impaired water bodies in the State of California's 303(d) list under the Clean Water Act for diazinon. Diazinon has been measured at levels that suppress or cause high mortalities for benthic insect life throughout the San Francisco Bay Area at the time that these restoration projects were being installed. This insecticide is now illegal for most uses because of its toxicity, and its use was canceled for residential use in 2004.

The Baxter Creek at Poinsett Avenue case indicated an improvement in its benthic assessment at a reach-scale restoration, only to have the project match background conditions with unrestored sections over time. Although the percent impervious may indicate biological challenges, should we also use that as the prime indicator for the return of loss of stream functions? Certainly these cases represent reach-scale projects that return riparian corridors, in-stream complexity, sediment transport and storage dynamics, and a desired channel stability that maintains itself dynamically under flood flows. These projects represent the case that even with high impermeability and high urban densities, it is possible to restore quasi-equilibrium channels and floodplains and stream processes, even if some of the functions are impaired by pollution.

The condition-based rapid biological assessments described in the Baxter Creek case do not record these processes and functions, and we are not investing in the widespread use of rapid assessments that can measure the incremental improvement of urban stream processes and functions. Codornices Creek in Berkeley, for example, with a computed watershed impervious cover of 34 percent, supports a remarkable steelhead population in a downstream refugia near the bay *before* any habitat enhancements were started. The East Bay Wildcat Creek watershed in Richmond, at 20 percent imperviousness, does have some unstable stream banks in the most dense urban corridor, but does that impervious classification predict that restoration efforts will not return the creek to some predevelopment stability or even predevelopment biodiversity? This question is difficult to answer because we are not positive, for example, of what predevelopment salmonid populations existed in Wildcat Creek, if any. Because the citizenry and agencies got involved with the creek in 1982, however, steelhead and rainbow trout were introduced to the creek, and, based on annual monitoring by the East Bay Regional Park District, those populations have maintained themselves for more than thirty years (Wise, Alexander, and Graul 2007; Sullivan 2015). Numerous stream reaches on Wildcat Creek, some more than a mile in length, have remained dynamically stable since 2000 and now feature a lush riparian corridor not existing over thirty years ago. This case raises the interesting issue that although Wildcat Creek will never return to predevelopment channel forms, the modifications made to the channel created a perennial channel in the lower reaches that may have a greater capacity for some ecological functions. The dense riparian forest and perennial presence of water and deep channel forms mean that this area is capable of supporting introduced salmonids, a function that this environment may not have provided historically.

Baxter Creek, at 65 percent imperviousness, certainly has relatively stable channels in urban backyards, and biological and water quality monitoring indicates good to fair water quality. The poor water quality designation based solely on a benthic assessment alone probably does not represent the actual range of water quality conditions. The Baxter Creek projects feature substantial riparian corridors and address previous stream disequilibrium at a number of project sites. Village Creek, with 55 percent imperviousness, has a channel daylighting project that surveys show has remained geomorphically stable over a fifteen-year period and supports a riparian corridor resembling a previous much less impacted landscape in the 1930s.

The population densities shown in table 4.1, an indicator of degree of urbanization, could indicate that it should be harder to restore a stream channel in the City of Berkeley than in Los Angeles because Berkeley has a greater population density. The range in percent impervious cover is very similar for the two cities. The population densities in the table are averages and vary in different parts of the two cities. Berkeley is nonetheless known for its unengineered, dynamic daylighting restoration projects. Los Angeles has been more constrained than Berkeley in enhancement or restoration because many of its streams and rivers are classified as flood control projects. The number of variables involved in enabling or constraining urban stream restoration projects has always struck me as too numerous to establish gross predictors of potential for restoration success or ecological recovery with a simple parameter such as imperviousness or population density.

Relatively recent research by fish biologists who were testing rapid assessments to rate the quality of aquatic environments found that comparing five different measurements commonly used to rate habitat, including aquatic macroinvertebrate indices, provided significantly different outcomes. Their assessments of the aquatic environment changed based on which index was used. Because each assessment provided a contributing evaluation to the environmental evaluation (presence of fish, amphibians, macroinvertebrates, physical habitat, and vegetation), they recommend combining a multimetric approach to represent a rapid assessment of environmental conditions as opposed to relying on any one of them (Aparicio et al. 2011; Purdy, Moyle, and Tate 2011).

The experience derived from the cases in this book is that the most basic limiting factor for establishing channel complexity and stability has been the widths of available rights-of-way to allow ecological functioning to return to floodplains and channels. Although population density and impervious cover can be *indicators* of

TABLE 4.1.

Estimates of imperviousness and population densities

	Imperviousness (%)	City	Population Density (per mi²)[a]
Wildcat Creek[b]	20	San Pablo	11,727
		Richmond	3,310
		Average population density	7,518
67% of watershed in East Bay Regional Park District and open space			
Baxter Creek[b]	65	El Cerrito	2,870
Codornices Creek[c]	34	Berkeley	9,823
Village Creek[c]	< 55	Berkeley	9,823
Blackberry and Strawberry Creeks	No estimate	Berkeley	9,823
Alhambra Creek[b]	15	Martinez	2,993
Los Angeles River[d]	47	Los Angeles	7,876

[a]US Geological Survey National Land Cover Data for population density.
[b]Contra Costa County 2003.
[c]2001 *Watershed Management Plan*, City of Berkeley Public Works Engineering Section.
[d]Los Angeles County Public Works Department. Listed to provide comparison with a known high-density city.

environmental decline or recovery, we are on shaky ground to use them as *predictors* of restoration potential or restabilizing streams. This awareness confirms the warnings found in the literature that the impervious indexes may not apply to all areas of the country and that many environments will not conform to their averages.

Channel Evolution School

The channel evolution school provides sketches representing watershed and channel processes that can offer very useful descriptive information about what may be causing stream imbalances or how the stream is reacting over time to watershed land uses. A limitation of these stream evolution models is that the stream may not follow the typical evolutions expected. Stabilizing bedrock or channel hardening may "arrest" channel evolution so that channels affected by channelization or other urban impacts remain entrenched U-shapes rather than evolving into more recovered systems. In contrast, the more recovered streams are located in entrenched valleys, but they form wide "inset" floodplains that add to channel stability and ecological, flood reduction, and erosion reduction. The advantage of channel evolution and process models is that they are particularly well suited for neighborhood-scale projects in which it can be hard to justify the expense of setting up complicated sediment transport models to predict channel responses. Probably the best practice in these cases is to create as much floodplain space as possible in project design as insurance to allow for future channel adjustments over time to any additional watershed land use changes (Thorne 1999; Cluer and Thorne 2014).

Channel evolution models were of use in these cases because they structured a mental process for understanding what the channel may have been historically, what was governing the past channel type, and what future tendencies for adjustments the channel may make. It helped us focus on the causes of channel instabilities that we were noticing. This school particularly came into play on the Glen Echo Creek project in which we knew that the channel was going through a modest widening and incising process. The design therefore provided a slightly wider active channel and set back terraces to anticipate future adjustments (although this early project is much overengineered). Reference reaches are viewed in the context of channel evolution in that nearby stream reaches were rejected as references if they did not represent sites that had adjusted to the current urbanized conditions, a deduction made after taking stream channel walks and gathering information from adjacent property owners or local agencies.

Generally, the widely used Simon and Schumm evolution models (Schumm, Harvey, and Watson 1984; Simon 1989)—which predict that channels will entrench when degraded by many land use impacts, followed by eroding, collapsing stream banks, channels widening, and the formation of new floodplains—were not applicable to many of our sites. Many cases described here are constructed in steeper coastal streams and often skip the stage of building inset floodplains from the reworking of collapsed banks because the bank sediment mostly gets transported fairly quickly as suspended sediment. The most relevant channel evolution models in these cases were the observations and recognition of an urbanization cycle in which the end stage is a stream system with more frequent and greater peak flows, a decrease in sediment load and wider channels, and larger cross-sectional areas. In an ironic twist, we were at an advantage to be working in an older, established developed urban area in which any additional infill development typically may have had only a localized impact, if any, without changing the urban equilibrium channels that formed over a long period of urbanization.

Stream Process Descriptions

The collection of qualitative and quantitative data on stream processes was an important aspect of evaluating the watershed conditions and how they would influence the reach-level projects with which we were involved. Observations of the watershed helped us ask where sediment sources were probably coming from, whether the stream system had a continuing instability because it was sediment limited, or whether the watershed had an imbalance between high sediment loads and discharges. Field observations also told us a great deal about the channel substrate, the channel banks and terraces, and the basic type of soils or bedrock present affecting the stability of the channel. Most of the lower-watershed project locations had structural soils composed of dense clays and silts. The middle reaches composed of gravel and cobble streambeds were typically stable unless straightened, hardened, or otherwise modified by property owners. Although it is safe to assume that most of these urban watersheds did not have as great a sediment load as preurban situations, a mix of sediment sizes is making its way down the watersheds systems from the upper areas that are characteristically unstable hillslopes located in parklands, which in many instances are parks because they are too unstable to develop. Channel slopes have been controlled over time by frequent culverts and bridges. Sediment, stored in point bars and behind hydraulic constrictions, pulses down in high flows. In no project reach did we find excessive limitation of bedload, nor were there sources of sediment producing excessive aggradation except in the case of Blackberry Creek. In that case, the best hypothesis from the city was that the excessive sediment was a result of broken water supply lines crossing creek channels. A well-regarded local historian, Richard Schwartz, found records of multiple, unregulated gravel and rock borrow areas extracted in the East Bay hills that add to upper watershed instability and may be part of a long-term episodic source of sediment entering some of the creek watersheds.

Using empirical data and observing small-scale projects gave us an opportunity to witness the adjustments and reactions of these stream systems after unlocking them from culverts and ditches. Starting projects small, observing the sediment transport regime, interviewing property owners, public works officials, and historians on their observations and knowledge, and keeping notes was a more reasonable approach than numerically modeling sediment transport.

Applying Analogs

Imagine the challenge of restoring a ditch or a culverted stream to a quasi- equilibrium condition in an urban watershed where a significant part of the stream may be underground or greatly modified by property owners. The first clues that we used to chart a course were historic maps to understand the creek's location and to see if this creek used to be a single-thread, multiple-thread, discontinuous, straight, or sinuous channel. Typically, the lower reaches of the channels in areas of tidal influence have been greatly modified by humans, and, given today's land use constraints, re-creating historic stream processes associated with wetland alluvial fans, distibutary channels, and forms involving widely meandering, multiple, looping channels is impossible. It was interesting to observe that some of the channels in the midportions of watersheds nonetheless still follow old channel locations and have retained similar sinuosities over time.

Channel types and meander lengths could be estimated by looking at historic maps. Our office collected historic maps from public libraries, the University of California Water Resources Center Archives, and the university's Bancroft library as well as old maps that people donated to our creek groups. Ranchero property and coastal surveys from the mid- to late 1800s were the staple of our office décor. Aerial photos starting from the 1930s and 1940s were available for many areas in the East Bay. These historic records were used in two ways. One was to understand what type of creek system used to be on the site where we were working. Village Creek, Baxter Creek at Booker T. Anderson Park, and Alhambra Creek were clearly influenced by the tides from San Francisco Bay. Village Creek and adjacent Codornices Creek appeared to be disconnected channels navigating across an alluvial fan entering a bay wetland. Historic aerial photos informed a sinuosity for a re-created, meandering single channel on Village Creek, even if the original disconnected wetland channel forms could not be re-created. Aerial photos as recent as the 1960s informed a reach-level project on Codornices at Sixth Street and Blackberry Creek, both in Berkeley.

The second need was to find an analog representing current watershed conditions that could inform the restoration dimensions for a creek that we wanted to daylight from a culvert or recover from ditching and straightening. In all the reach-scale cases, there was at least one nearby site, sometimes found in stream segments in backyards, that could give reasonable information on stable channel dimensions. How do we determine if this reach is a reasonable reference for a stable channel under current conditions?

Empirical Methods, Including Hydraulic Geometry and Regional Curves

The tool we went to first for determining if some reaches were candidates for a reference reach was the San Francisco Bay Area regional curves data developed by Luna Leopold when he arrived in the Bay Area in the 1970s. The Bay Area regional curve provided our first guidance on what average channel dimensions and channel-forming discharges could be for the drainage area under consideration (Dunne and Leopold 1978; Leopold 1994). The regional curve cross-sectional area and the channel widths based on drainage areas provided a starting point for identifying potential reference sites that had dimensions close to the regional average. Numerous cross-sectional surveys were then plotted for candidate reference reaches that were at the same slope and had similar soil and geologic conditions. If the surveys indicated that we were getting consistent values 80 percent of the time or more for active channel cross-sectional areas and widths, we proceeded to use them as reference sites. A profile would then be surveyed at estimated floodplain levels through the cross sections to further confirm a floodplain-active channel elevation. The Bay Area is characterized by progressively incising channels, which the cross-sectional surveys record. We knew that it was important not to mistake one of the several higher abandoned terraces as the active channel and selected for the lowest-elevation grade breaks to designate floodplain elevations in field investigations.

Monitoring of our first restoration projects indicated that the design channel widths were filling in after the first channel-forming flows and that depths remained close to the original project design or were slightly less. Post-project surveys at five project sites indicated that cross-sectional areas were measuring a consistent 30 percent less than the designs. As a result, it seemed prudent to collect data to modify the San Francisco Bay regional curve to better represent the channel geometries for drainage areas for the East Bay subregion. Using the adjusted channel dimensions from post-project monitoring and data collected from four gage stations located in the East Bay, the Waterways Restoration Institute developed an East Bay regional curve. The curve used data from nine sites, which is not an ideal number of data points, but it nonetheless improved our project design curves and was applied to the projects we designed starting in 1999. The slopes of the curves plotted parallel to the 1978 published Leopold curves and the plots were consistently below the Bay Area curve. The two curves are shown in figure 4.1. Upon consulting with Leopold on these findings, he found it to be perfectly logical that the East Bay characterized by an average annual rainfall of 22 to 25 inches should plot below the values computed in his Bay Area curve, which represented an average annual rainfall of 30 inches (fig. 4.1).

These reach-level projects did not have the benefit of data from hydrology gaging stations. The best hydrologic information is, of course, obtained by stream gages maintained by the US Geological Survey (USGS) or by states, counties, or local districts that measure discharges, velocities, stages, and cross-sectional areas. The USGS produced a report to guide the estimation of hydrology for ungaged watersheds in the San Francisco Bay Area that employs a multiple regression analysis to correlate a range of flood discharges with selected watershed characteristics (Rantz 1971). This report derived its analysis from flood frequency relations from forty gage stations in the nine Bay Area counties. Some county public works departments have prepared flood frequency analysis based on rain gage and flow data and apply rainfall intensity data to derive flood magnitude recurrence intervals. Blackberry Creek hydrology was aided by a city engineer who had a gage in her nearby backyard and measured rainfall intensities. Table 4.2 indicates the estimated flood flows experienced by projects in the East Bay. The Bay Area

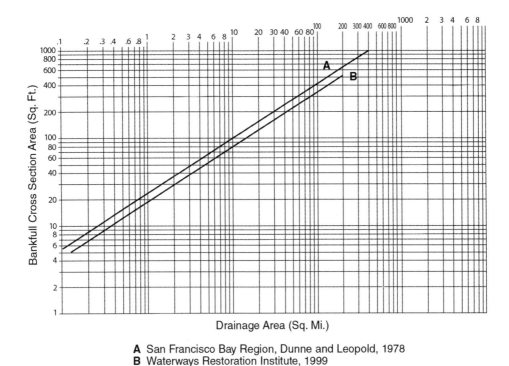

A San Francisco Bay Region, Dunne and Leopold, 1978
B Waterways Restoration Institute, 1999

Figure 4.1 The East San Francisco Bay curve was plotted to represent average channel dimensions for drainage areas better than the San Francisco Bay Area curve does. Credit for the San Francisco Bay Area curve: Dunne and Leopold 1978.

regional curve developed by Leopold provides a good cross-check for the USGS regression values for the two-year recurrence interval estimate. Leopold estimated that bankfull velocities for Bay Area streams tend to between 5 to 6 feet per second. Using this value multiplied by a cross-sectional area can also provide a back-of-the-envelope check for estimated discharges (table 4.2).

A variety of cross-checking computations were made to derive channel-forming discharges and dimensions and to estimate flood depths for different recurrence intervals. Dimensionless rating curves and other useful relationships between watersheds and discharges were published in Leopold's 1994 book, *A View of the River*, which he wrote to help restorationists. The relationships are correlated with different regions of California and the United States. Dimensionless rating curves can be used to relate average depths of the ten-year, twenty-five-year, and fifty-year recurrence interval flows to bankfull depths. Regional relationships have been computed between the average annual discharges and bankfull discharges. Using a value of 10.8 cfs as an average annual discharge per square mile of watershed and the equation that the average annual discharge is about 0.034 of the bankfull

TABLE 4.2.

Estimated flood flows experienced by projects (recurrence interval in years)

	1986	1995	1997	1998	2006
Blackberry Creek		1 in 10	1 in 25		1 in 25
Baxter Creek			1 in 25		1 in 25
Village Creek					1 in 25
Codornices Creek at Sixth Street			1 in 25		
Wildcat Creek	1 in 6	1 in 10	1 in 15	1 in 10	1 in 21

discharge could provide a check on whether channel-forming discharge estimates were reasonable (Leopold 1994). Some USGS water supply papers publish flood frequency curves and other flood analysis. The water supply papers publish records on annual floods that are the average of the maximum peak flood for each year of record, and they generally compute as the 2.33-year recurrence interval flood. Reach-level projects on Wildcat Creek, for example, benefited by years of gage data, and we developed flood frequency curves from the gage data. We started with the values for the 1.5-year recurrence interval flood, an average value representing the channel-forming discharge that was later adjusted higher after combining this with effective discharge computations.

We also applied some of the basic hydraulic geometry science developed to design channel lengths and sinuosity (Leopold, Wolman, and Miller 1964). Figure 4.2 illustrates the geometry of a meander. Other equations were used to calculate the meander length of a stream channel based on channel width and the amplitude and radius of curvature of a meander (Leopold and Wolman 1960). Using these three parameters, we drew a theoretical meander on paper and measured the length of the channel.

The equations are the product of averaging data from many rivers, and there is, of course, a natural variation that deviates from these numbers to represent a broader range of values. Generally, the guidance that Leopold gave for restoration designers is that average channel meander lengths are between ten and fourteen times the channel widths. The radius of curvature of the central portion of a channel bend averages about one-fifth of the meander length and is commonly two to three times the channel width. Pools and riffles are spaced at repeating distances on the average of five to seven channel widths (Leopold 1994). Leopold warned sternly not to use these equations to draw perfectly shaped sine curves and then excavate these uniform forms on the landscape. How we used the values for channel length and amplitude was to get this information into the planning process so as to get enough floodplain space to accommodate a channel length and meander belt (amplitude plus channel width) that was going to add to the stability of the

GEOMETRY OF MEANDERS

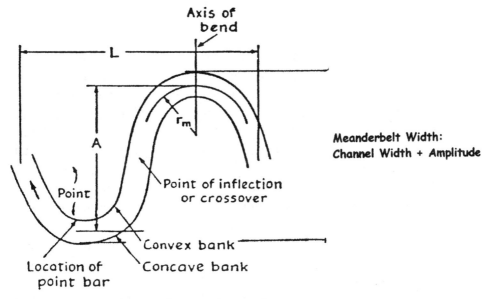

L = Meander length (wave length)
A = Amplitude
r_m = Mean radius of curvature

FIGURE 4.2 The geometry of a meander includes the length, amplitude, and radius of curvature. Credit: Leopold and Wolman 1960. Meander Length to Channel Width: L = Meander length (wave length) W = Width $L = 10.9w^{1.01}$ (The meander length ranges from 10–14 times the channel width) Amplitude of Meander to Channel Widths: A = Amplitude W = Width $A = 2.7w^{1.1}$ Meander Lengths to Radius of Curvature: L = Meander length r_m = Mean radius of curvature $L = 4.7r_m^{0.98}$ (The radius of curvature averages about 1/5 of the meander length and approximately 2.3 times the width)

design. As described in chapter 3, to create a more random meander design for Village Creek, I dropped a piece of string representing the proper channel length onto the paper design plan and produced a meander shape by randomly fitting all the string length onto the allotted right-of-way length and width. If my string created a radius of curvature exceeding the ranges by a significant amount as the string fell on the paper, I slightly reshaped the curve.

When viewing our historic maps from the 1800s, it was interesting to note that the streams tended to have fairly tight meanders and quite a bit of sinuosity in the mid-portions of the watersheds. The East Bay Wildcat and San Pablo Creeks' 1880 surveys, for example, indicated sinuosity ranging from 1.4 up to 1.9 in the lower to mid-portions of the watersheds before steeper terrain reduced the sinuosity to 1.3 and less. As the creeks entered the bay, more multiple-stem channels can be seen as well as meanders that become longer and amplitudes that increase the size of the meander loops. Unfortunately, we do not have good information on the channel widths from these maps to better understand the historic width and meander length relationships for either the fluvial channels or the tidally influenced stream types. For tidally influenced areas, the tidal prism drainage areas are substituted for watershed drainage areas to derive regional hydraulic geometry relations for channel widths, depths, and cross-sectional areas. A dearth of reference sites to understand how channel shapes change for fluvial channels transitioning to tidal channels creates a project design challenge if we are trying to help these channels recover from ditching.

The Oakland Museum maps that recorded existing watershed conditions and identified where unchannelized and unmodified channels are located reinforced that the average values computed by Leopold and Wolman (1960) for channel geometry relations were relevant for our region in current times. It does appear that some of the wider urbanized channels have lengthened channel meanders and reduced sinuosity in some areas of the bay. The current relationships between channel widths and meander lengths is a complex situation in which some of the loss of sinuosity is due to land use encroachments on the channels; therefore, we cannot necessarily equate the current meander patterns to changes in watershed discharges or sediment conditions and channel widening. We experimented with the idea that we could develop different average ranges for meander lengths for different channel types. For example, the flatter, tidally influenced channels could more typically have channel lengths approximately twelve times the channel widths and greater average values for amplitudes; the upstream meandering channels with well-developed floodplains could typically be about eleven times the width; and mid-slope channels that are confined between terraces but have sinuous channels with tighter bends could have average meander lengths closer to ten to eleven times the width. Our design methods mostly selected the value of between ten and eleven channel widths to draw theoretical channel shapes. Unfortunately, we never had the resources to develop a regionalized set of data to draw relationships between channel types and meander dimensions, but we hope there is now greater interest by government agencies to invest in developing this information.

By 2005, the formation of the Bay Area Watershed Network brought together a mix of environmental interests that up to then had been slow to organize themselves in the Bay Area. The network includes a mix of restoration design companies, local planners, watershed councils, and professionals in local and regional public works agencies. The list of priority needs for assisting their work included developing a better awareness of the range of assessments that can be applied to solving watershed management issues and needs. The needs list included developing regional curves that reflect the differences in the Bay Area's subregional environments.

In 2009, the Santa Clara Valley Water District developed a regional curve for the south bay for Santa Clara County that represented an average annual rainfall of about 14 inches (figure 4.3; Lee 2012). The San Francisco Estuary Partnership and the EPA supported the development of regional curves for north bay streams using data from fifty-seven Sonoma and Marin Counties stream sites shown in figures 4.4 and 4.5 (Collins and Leventhal 2013). The combined average rainfall for these north bay sites was 37 inches. This north bay data collection diverged from the original Leopold Bay Area curve in that a greater range of channel types were surveyed and more data were collected representing smaller drainage areas. Note that the slopes of the curves deviate from the original Leopold curve and more accurately represent the conditions found in this subregion using data from more stream types.

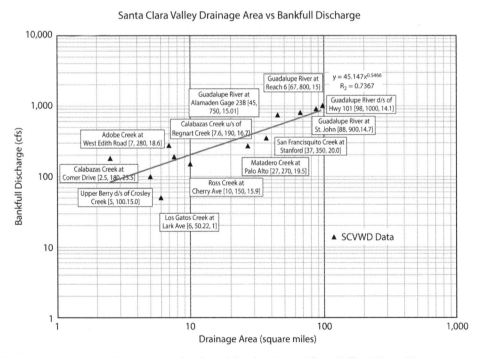

FIGURE 4.3 A regional curve was developed by the Santa Clara Valley Water District in 2009 using gage data from San Francisco Bay south bay gaging stations. The curve represents an average annual rainfall of 14 inches. *Credit: Lee 2012.*

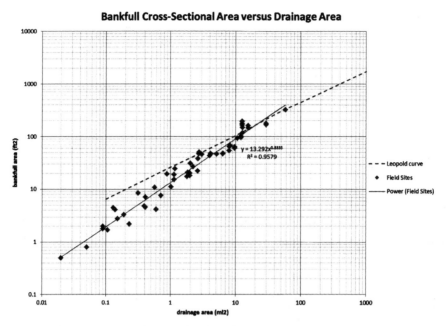

FIGURE 4.4 The 2013 regional curve developed for San Francisco Bay Area north bay counties shows cross-sectional area versus drainage area. *Credit: Collins and Leventhal 2013.*

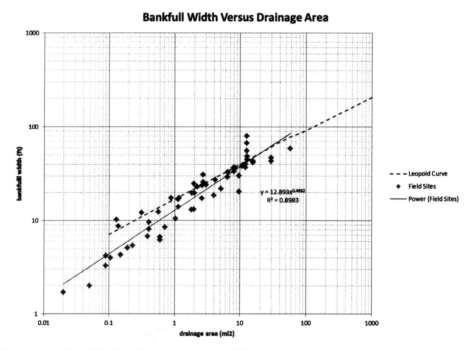

FIGURE 4.5 The 2013 San Francisco Bay north bay curve shows relationships between bankfull widths and drainage areas. *Credit: Collins and Leventhal 2013.*

A next phase of analysis for the north bay curves will segregate the regional curves by more stream types (pool riffle, step pool, tidally influenced, etc.). An interesting feature of this work is the relation developed between the upstream drainage network length and the bankfull cross-sectional areas. The network length included channel lengths, ditches, and culverted sections above the area selected for data collection sites and was found to be a very good predictor of bankfull cross-sectional areas shown in figure 4.6. Drainage network lengths, taking into account storm drains, could become a better predictor of hydraulic geometry for urban areas than simply measuring a watershed drainage area (Collins and Leventhal 2013). Figure 4.7 illustrates a project by San Mateo County on the peninsula below San Francisco (San Mateo Countywide Water Pollution Prevention Program 2013) to combine a number of subregional curves from different geographic areas from the greater Bay Area.

The post-project surveys of the projects described in chapter 3 indicate that the application of regional curves for designing channel dimensions was a very useful design tool even in a heavily urbanized environment. As mentioned, we were careful to cross-check the regional curve values against estimates of channel-forming discharges and watershed hydrology and nearby references. In a few cases, we cross-checked against HEC-RAS model outputs. The restoration channels adjusted over time but remained close to the design cross-sectional areas. The adjustment we made in the use of this tool was to realize that the large number of

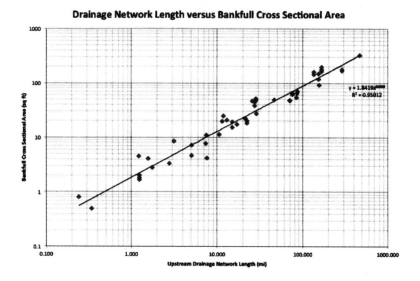

FIGURE 4.6 The 2013 San Francisco Bay north bay curve shows strong relationships between a bankfull cross-sectional area and channel network length. *Credit: Collins and Leventhal 2013.*

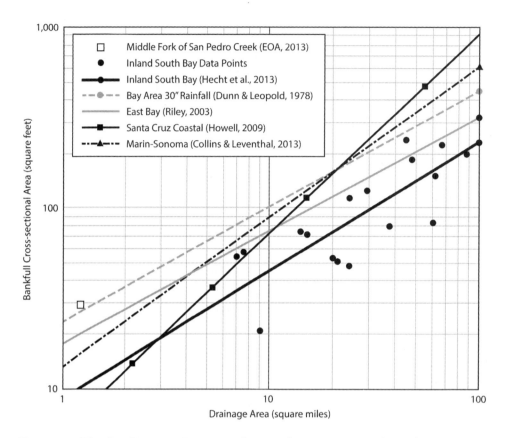

FIGURE 4.7 The San Francisco Bay area subregional curves compared. Credit: San Mateo Countywide Water Pollution Prevention Program 2013.

different microclimates in the Bay Area, with significant shifts in annual average rainfall within just a few miles, required the development of San Francisco Bay Area "subregional" curves to represent the climatic differences. The other lesson we learned with the use of regional curves was that although the flat-gradient streams near San Francisco Bay may exhibit average cross-sectional dimensions based on watershed drainage areas, the distribution of the channel area between widths and depths did not conform to regional averages. These flat-gradient stream types therefore adjusted to much lower width-to-depth ratios, reinforcing the need to begin to segregate the collected field data by stream types.

Design using hydraulic geometry relations between channel widths and lengths along with historic data seems to have guided the creation of more stable planforms based on a twenty- to thirty-year project monitoring record. In situations in which channel lengths were added to ditched systems, planform erosion problems are dramatically reduced or are no longer an issue. Surveys over time generally indicated some increases in channel sinuosity over design dimensions, and there have not been channel "cutoffs" indicating that channel slopes were overly flattened by the designs.

Some professionals involved in the river restoration design field caution against the application of regional curves for restoration design in urban areas because the curves can assume that the channel dimensions are formed by an average recurrence interval of 1.5 years, even though the channels may likely be formed by more or less frequent discharges. The incised nature of urban streams can make field surveys of bankfull dimensions difficult to distinguish (Copeland et al. 2001; National Resources Conservation Service 2007b). Cognizant of these issues, we first encouraged only the use of seasoned professionals to develop regional curves, so the scientists who developed the curves in figures 4.1 to 4.6 were particularly qualified to use good professional judgment on the use of both gage and field data to develop them. The field scientists understood the need to carefully select reference sites and not make rigid assumptions about the recurrence intervals of channel-forming discharges. If gage data represent a long period of record for fully urbanized land use and the channels represent an enlarged urban cross section, the 1.0- to 2.5-year recurrence interval range for estimating channel-forming discharges is a reasonable value to begin an investigation of bankfull discharges as suggested by Leopold (1994). In one East Bay restoration case in which a long period of gage data is available, an effective discharge was computed by the US Army Corps of Engineers for Wildcat Creek to determine which discharge over a period of time transports the most sediment, another expression of channel-forming discharge. This computed discharge for channel-forming flows changed our starting assumption of a 1.5-year recurrence interval bankfull flow to about a 2.8-recurrence interval, or from 300 cfs to 500 cfs. Wildcat Creek has an unusually high sediment load, which would explain the higher magnitude of flows needed to transport the sediment loads.

Numerous urban and rural streams cannot produce field data for equilibrium conditions, and in these situations, it is more appropriate to apply dimensionless hydraulic geometry relations computed from watersheds in the region. A rating curve of discharge versus depths can be developed in dimensionless terms that represent an average stream channel in a region. Regional differences tend to be small, so a general curve will give a good approximation of the relationship of discharges to depths for an ungaged watershed (Leopold 1994).

The additional reason hydraulic geometry relations can succeed in providing good design guidance for these projects is that the East Bay has been in a built-out urbanized condition for thirty to sixty years, and the streams represent an "urban equilibrium" condition adjusted to watershed conditions that are not experiencing changing hydrology or sediment budgets. This consideration is important for rural areas as well. A report on regional curve development for the rural Catskills Mountains in New York notes that many streams in their regional curve data sets include reaches that were previously or are even presently located in a disturbed watershed, and it found that the consistency and reliability of results was more dependent on the degree to which a site has been functioning under the current hydrologic regime (Miller and Davis 2002).

Analytical School and Hydrology

Analytical methods are based on physical equations, which can account for sediment inputs. They provide unique deterministic solutions, and because they apply basic physics of flowing water and fluid mechanics, they elicit the confidence associated with an older, established science. This school has a historic advantage in that it represents conventional engineering approaches accepted since the 1930s. On the other hand, the wide application of these tools to channelize rivers and attempt to simplify river environments became the motivation to rethink and improve analytical tools and their use (Soar and Thorne 2001).

Most restoration practitioners use at least some of the three governing equations of continuity, flow resistance, and sediment transport because of their practical value. These hydraulic relationships recognize that the stream discharge is the product of cross-sectional area, flow, and velocity; that velocity is a function of stream depth, slope, and channel roughness; and that sediment transport is a function of sediment sizes, stream discharges, and channel roughness. The analytical methods, also sometimes referred to as the rational methods, represent a mechanistic attempt to balance the forces from fluid motion of stream flows with the resisting forces of mobile bed and banks so as to design stabile dimensions for streams.

Hydraulic analyses are used in restoration design to model water-surface profiles or elevations through a design reach and can be used to predict flooding over the banks of streams or terraces under different design scenarios. The models can be used to estimate velocity distributions for evaluating habitat features and can become part of fish passage analysis. Analytical analysis includes two forms of sediment transport analysis: incipient motion analysis and sediment discharge analysis. The incipient motion analysis is used to estimate the maximum size of streambed particles that can be picked up and transported at a selected discharge. This calculation is useful for determining erosion or sedimentation potentials of channel designs. Incipient motion analysis can be adequate for many design situations where bed material is not excessive and sediment discharge imbalances do not appear to be a concern. Sediment discharge analysis is done when there is an issue that sediment loads may not be in balance with discharges. Sediment transport analysis can require a substantial amount of field data on channel dimensions and flow characteristics (Thorne, Hey, and Newson 1997; National Resources Conservation Service 2007b; Skidmore et al. 2009).

The channel continuity equation (also called the conservation of mass equation) forms the basis for most beginning hydrology and hydraulics instruction and establishes that stream discharges are the product of the cross-sectional areas of stream channels and velocities. Flow resistance can be calculated using a number of equations, with Manning's equation the best known and most widely used. The parameters in this equation can include values for channel roughness, discharge, mean depth, and slope, and it is used to solve for values of discharge, velocity, and roughness. Where channel dimensions and flow discharges are known, the roughness coefficient can be calculated. In most circumstances, however, this value is simply estimated, and the value selected can greatly influence the results of calculations and hydraulic models.

Because the analytical equations do not capture all the variables acting on a stream system, efforts have focused on how to use process-based methods so that they can better capture more natural processes or be combined with other methods to represent a wider range of variables that can better address restoration design. A common expression of discomfort with the empirical school is its reliance on the collection and analysis of field data, which can cause errors along the process. Although founded in physics, many hydraulic relationships likewise require empirical coefficients to represent estimated values involved in the river processes.

In these neighborhood-scale projects, the analytical tools used were mostly Manning's equation, critical shear stress computations, and shear-stress-based computations, which were mostly applied to checking channel dimensions first selected through empirical and analog methods. Figure 4.8 contains these equations. In most cases, sediment transport models and hydraulic water-surface elevation models were too expensive to use, and in some cases, appropriate levels of gage-based discharge data or sediment data did not exist, so it would be contrived at best to attempt to quantify the sediment transport regime for these small watersheds in the context of these reach-scale projects.

That said, we understood that sediment transport had to be a critical consideration in any stream project design. The only creek in our area of practice in which high sediment loads per acre of watershed created a concern about the competency of the stream was Wildcat Creek, and the complexity of that project does not translate to these reach-scale projects. Our best sources of information to determine the transport competency of design proposals ended up being empirical and analog data derived from a number of monitored nearby restoration projects that made consistent adjustments to restoration projects. In some cases, shear

Manning's Equation

Solving For Discharge:

$$Q = \frac{1.49 \ (A) \ (R) \ 2/3 \ (S) \ \frac{1}{2}}{N}$$

Solving for Velocity:

$$V = \frac{1.49(\ R \ 2/3 \ ) \ (S1/2)}{N}$$

Solving for Roughness:

$$N = \frac{1.49}{(Q) \ (A) \ (R \ 2/3) \ (\ S \ 1/2)}$$

A = cross-sectional area (sq. ft.)
R = hydraulic radius (mean depth)
S = channel slope
V = velocity (fps)
N = roughness

<u>**Shear Stress**</u> (or tractive force in pounds per square foot)

$$T = (Y) \ (\ S) \ (R)$$

Y = specific weight (density) of water or (62.42 pounds per cubic foot)
S = slope
R = hydraulic radius or mean depth

Critical Shear Stress (or critical tractive force)

$$Tc = T^*(Ys\text{-}Y)D$$

Tc = critical shear stress (pounds per sq. foot)
T^* = dimensionless Shields parameter
Ys = specific gravity of sediment *(165 lb/ft3)*
Y = *specific weight of water (62.42 lbs/ft)*
D = sediment particle diameter (ft)

This Shields Critical Shear Stress Equation (1936) obtained values for the T^* experimentally using uniform bed materials and measuring sediment transport over them at decreasing levels of bed shear stress. (For example, a .5 dimensionless value is appropriate for clay soils and .06 for gravels and cobbles) refer to Fischenich, Craig, May 2001, "Stability Thresholds For Stream Restoration Materials," U.S. Army Corps of Engineers, Research and Development Center

FIGURE 4.8 These analytical equations are commonly used in restoration design.

stress conditions were calculated and compared to the computations of critical shear stress needed to start entrainment of dominant particle sizes to provide a useful check on the draft designs for equilibrium dimensions first developed through hydraulic geometry and reference information. A reach-scale project designed for Wildcat Creek at Davis Park became a good example in which relying on only an analytical approach without checks from empirical data would likely have created a sediment transport problem because the shear stress analysis did not take into account the high sediment loading of this watershed. Occasionally, we used a calculation of a friction factor, which is a relationship between mean velocity and shear velocity (a theoretical velocity), for estimating a roughness value as a check against Manning's equation.

The HEC-RAS water surface models were used in a few of the reach-level cases, more as a legal exercise for the cities sponsoring the projects to verify flood protection objectives as opposed to a project design tool. In some situations, projects were located in chronic flood-risk areas, and local agencies believed that they needed to run the model to establish a record that the flood hazard or water-surface elevation was being reduced and not increased for neighborhoods adjacent to the restoration project. Because Village Creek was part of a development project that needed to attain flood protection for the housing development, the engineering firm involved with the project checked the water-surface elevations of the restoration design of the estimated one-hundred-year discharge with the HEC-RAS model. The City of El Cerrito wanted to indicate that there would be flood reduction benefits for the Gateway project on Baxter Creek. Shear stress calculations using the equation in figure 4.8 proved useful in estimating the likely capacities for soil bioengineering systems to perform. After most of these projects were completed, the very useful permissible shear stress and velocity tables were published by the US Army Corps of Engineers and Natural Resources Conservation Service, providing numerical guidance for selection of soil bioengineering systems, which did not inform most of these project designs (Fischenich 2001; National Resources Conservation Service 2007a). The soil bioengineering systems selected for use at the project sites, including brush layering, brush matting, poles, and stakes, performed excellent stream stabilization services under the various flooding levels shown in table 4.2.

Stormwater models such as the widely used EPA Storm Water Management Model can produce estimates of discharges for the more frequent, lower-magnitude floods, although they consistently produced conservative (higher values for discharges) than these other sources. Table 4.2 represents the use of this full range of sources for hydrologic data. Although all these channels are located in a common region of the East Bay, the Bay Area is famous for its many microclimates and variations in rainfall within fairly short distances. The gage at Wildcat Creek does help inform regional flood recurrence intervals, but more localized information needs to be applied to the creeks located in other nearby watersheds. To check field results used to initially select restoration active channel dimensions, we entered values for slopes, roughness, cross-sectional area, and hydraulic radius (mean depth) into Manning's equation to estimate a discharge. The result was compared to the values we calculated for the two-year recurrence interval hydrology using regional regression information provided by the USGS. These discharges provided a check to determine if the selected channel dimensions appeared reasonable for active channel dimensions. Channel slopes were computed for restoring ditched, straightened, and culverted streams by drawing a meander based on the empirical equations in figure 4.2 and measuring the sinuosity produced. A valley slope was measured from site surveys, and the channel slope was computed by dividing the valley slope by the sinuosity, as illustrated in figure 4.9.

Shear stress calculations were used both to provide a more realistic picture of erosion potential as well as select an active channel dimension in which shear stress and critical shear stress values indicate design channel dimensions that are capable of entraining the dominant bed particles. Our process for accomplishing this step typically involved developing a value for the D84 particle size on the streambed (meaning that 84 percent of the particle sizes are finer than this size), which is associated with the size transported by the active channel discharges. The particle size distribution for a stream can be found through a Wolman pebble count using one hundred random collections across a streambed and, by plotting this information on a graph, can show the cumulative particle sizes versus the grain sizes. A value for critical shear stress can be computed from the D84 particle size using figure 4.10, which plots the critical shear stress at the threshold of motion for different grain sizes (Leopold, Wolman, and Miller 1964). The value representing the critical shear stress needed for entrainment of this particle size can be compared against the shear stresses acting on the channel. If the two values are close, the selected design active channel depth should not, theoretically, experience excessive erosion or deposition.

Biological Sciences

BIRD POPULATION STUDY

Studies show that birds are drawn to cities due to the availability of food, water, and irrigated landscapes. Researchers are recording an increasing variety of animal as well as bird species moving into cities out of the necessities created by losing habitat elsewhere. They are also recording the ability of bird and other wildlife species to adapt to and even genetically evolve to better occupy urban environments.

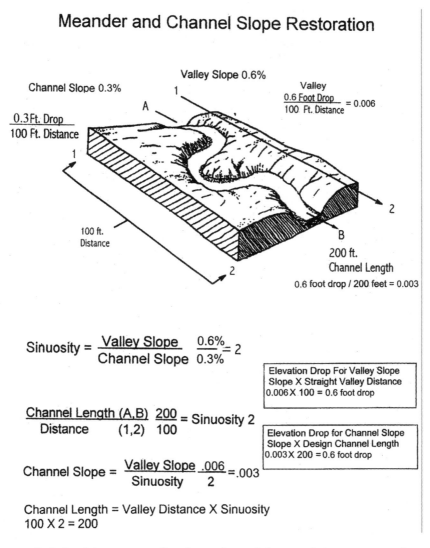

FIGURE 4.9 Relationships among valley slope, channel slope, and sinuosity can inform design.

There is an increasing record of urban areas functioning as refugia for endangered species (such as the peregrine falcons I watch from my fifteenth-floor office building in downtown Oakland) and therefore a resulting emphasis on what humans can do to modify urban environments to accommodate this migration to cities. In fact, ecologists are now stating that if we are going to conserve the diversity of species we have now into the future, we are going to have to work in urban habitats to meet this need (DeWeerdt 2014).

My earliest memories of professionals organizing themselves to develop a strat-

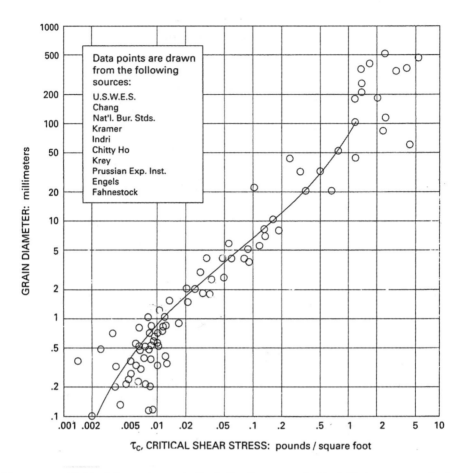

FIGURE 4.10 Critical shear stresses indicate the forces needed to initiate movement of streambed materials. *Credit: Leopold, Wolman, and Miller 1964.*

egy to advance the protection and restoration of riparian systems in California in the 1970s and 1980s was of the bird advocates who realized the large-scale disappearance of bird habitat associated with the loss of large-scale river riparian environments (Warner and Hendrix 1984). Just as fish biologists have developed recovery plans for native fish species, bird biologists have put together interagency and interdisciplinary teams to develop regional and subregional bird habitat recovery efforts. The bird biologists have strong ecological arguments for conserving birds as a featured component of biodiversity because of the critical roles that birds can play in ecological systems. By managing for a diversity of birds, other elements of biodiversity can be assisted and their presence, or lack of, is a sensitive indicator of environmental conditions. The economic value of fish and their link to the food industry has generally put this species at the forefront of attention. The bird-watching and ecotourism market, however, is a significant contributor to the economy. It involves 67.8 million bird watchers throughout the United States, and it contributes 16,000 jobs and $622 million annually in retail sales in California alone (Riparian Habitat Joint Venture 2004).

Planners devising a statewide recovery plan for riparian birds in California have identified four priority riparian restoration activities and locations: restore the natural hydrology of an area; locate restoration sites within the potential dispersal range of existing "source" populations of birds; protect adjacent upland habitat areas for foraging, food, and nesting habitat; and make land use improvements within the watershed to the extent practical. Bird ecologists echo the recurring theme that restoration tends to occur at reach scales, and good conservation biology needs to look at landscape connectivity and consider the wider ecology of landscapes where remnants are located. Bird biologists are interested in large-scale environmental features such as regional topography, urbanization, and forest fragmentation. The important landscape-scale factors that affect habitat are identified as altered hydrology, fragmentation through urbanization, rangeland, and forestry land use activities. The habitat fragmentation can affect the success of small- as well as large-scale projects. Although it is optimum to be able to restore or protect large continuous blocks of continuous habitat, the bird literature finds value in restoring patches of habitat and then working to reconnecting the patches to obtain the desired long-term population stability by providing for the interconnectivity of the fragments. Removing barriers to movement such as reducing distances between the fragments becomes an important strategy. As noted in the *Riparian Bird Conservation Plan,* "It is increasingly recognized that viewing habitat remnants as islands embedded in a sea of unsuitable habitat is an oversimplification of reality," and conservation planning needs to expand the model it is using to plan for phased, incremental restoration (Riparian Habitat Joint Venture 2004).

The habitat recovery approach also employs the concept of identifying so-called focal species to plan population monitoring and recovery efforts. The National Partners in Flight (NPF) program has motivated regional conservation planning and recovery efforts nationwide for birds and in California where the cases in this book are located. The NPF formed the interagency and nonprofit partnership, the Riparian Habitat Joint Venture. This California effort overwhelmingly selected riparian habitats as a top priority for recovery because they provide the richest habitats for both breeding and wintering birds. The Riparian Habitat Joint Venture developed a suite of focal species based on their ecological associations and conservation concern and their ability to indicate habitat characteristics and "healthy" riparian systems (Riparian Habitat Joint Venture 2004).

The joint venture selected species to monitor with distributions large enough to provide sufficient sample sizes for statistical analysis across sites and regions. They used these factors to select the species to guide planning recovery and monitoring: riparian habitat is the primary breeding habitat for the species in most biore-gions of the state; they warrant special management status as federal- or state-listed species under the Endangered Species Act or species of special concern; they commonly breed throughout California riparian areas; and they have breeding requirements that represent the full range of successional stages of riparian eco-systems. Seventeen focal species were selected for California. Species inventories are then performed to assess the success of restoration efforts.

Five East Bay streams were subject to a bird population study using the focal species assessment approach, which included chapter 3's Baxter Creek restoration sites at Poinsett Avenue and Booker T. Anderson Park. Three additional nearby restoration projects included in the study were the Wildcat Creek flood risk reduc-tion project area located near Verde School in unincorporated North Richmond, Wildcat Creek at Rumrill Road in the City of San Pablo, and Lower Rheem Creek at Contra Costa College, also in San Pablo.

Goals of the research were to examine bird response to riparian restoration in an urban setting and to identify factors such as restoration area size, time since restoration, surrounding land use, and habitat structure that could benefit bird populations. Bird monitoring was conducted by the San Francisco Bay Bird Ob-servatory using bimonthly surveys from spring to early fall for two years, 2009 to 2010. The research method used was to compare five selected riparian restora-tion sites against five control sites that were categorized as unrestored and against nearby remnant sites (Demers and Scullen 2010).

For each site, the researchers calculated total bird abundance (numbers per site), total bird density (birds per acre), species richness (numbers of species per site), and species diversity (index comparing the portion of a number of a certain species relative to all the individuals). Two years of data identified a total of 3,164 birds representing sixty-three unique species. These findings alone suggest sig-nificant bird use in densely populated urban areas, something that research has corroborated over years of study.

It was not surprising that one of the control sites with an older, mature wood-land located on upper Wildcat Creek in Alverado Park, which is part of a large regional park, had the greatest mean species richness and cumulative species rich-ness. The lower Wildcat Creek site at Verde School contained within 1-mile-long restoration and flood risk reduction project exhibited the greatest species diversity and second greatest mean species richness and cumulative species richness. Bax-ter Creek, at Poinsett Avenue and located in an urban setting between two streets, had the greatest bird density of all the sites. The state-threatened willow flycatcher was found unique to the Wildcat Creek site at Rumrill Road. A violet green swal-low was found unique to Baxter Creek at Poinsett Avenue. The two lower Wildcat Creek sites, which are almost contiguous, together contained the USGS-classified riparian obligate species, the willow flycatcher, and the orange-crowned warbler. The Riparian Joint Ventures focal species found at the restoration sites were black-headed grosbeaks, song sparrows, warbling vireos, yellow warblers, and Wilson warblers as well as willow flycatchers. Some of the sites classified as unrestored shared these same species, and two unrestored sites contained an obligate and focal species that the restored sites did not. For this reason, the authors concluded that this particular study did not differentiate between restored and unrestored sites.

The researchers struggled with a number of confounding variables in trying to discern relationships between populations and land uses. They concluded that the size of the site and the surrounding land uses had more influence on the population study than whether a site had experienced restoration. The authors found that their observations likely conformed with other urban bird population research findings with respect to older, more mature and larger riparian sites having greater species diversity. Increasing species richness and diversity as a function of riparian corridor age can logically be equated with the greater habitat complexity that comes with more vegetative layers and heights (Demers and Sculler 2010).

The riparian habitat joint ventures population assessment methods stress the use of indices of breeding obligate or focal species as the indicator of riparian health. This study of urban creeks found six riparian obligate species and seven joint venture focal species, many of which overlap, so that a total of nine obligate-focal species were found in the urban sites. Most of the birds observed were not in breeding pairs. According to the authors of this study, the relative lack of spotting breeding pairs does not necessarily indicate that the sites do not have the potential to support breeding pairs. They emphasize the value of the sites in functioning as migratory stopover habitats and supporting postbreeding habitat as well as providing movement corridors for all wildlife.

This research is an interesting starting point for studying bird populations in urban creek restoration areas in terms of reaffirming significant numbers of riparian birds in cities and helping inform ways to improve future research projects. Future urban bird population research efforts focused on restoration can be improved by clearly describing the definitions used for both restoration sites and nonrestored sites. In this case, the bird count in the unrestored sites was confused by including a site that should have had optimum habitat because of its mature woodland and location in a regional park system and did not seem to fit the definition that the control sites were located where "remnant" riparian areas were located. (That is not to say that an interesting research question is to ask how restoration sites compare with older mature wildlands locations, but that does not appear to be the intent of this research.) The definition issue arises again with inclusion of the Upper Rheem Creek as a "restoration" site, which may also skew research results because it does not fit the definition of restoration that we apply in this book. Although this research shows that similar obligate-focal birds could be found in both categories of sites, it does not provide a cumulative scoring of how many of the obligate bird species are found in each category, restored and unrestored.

Table 4.3 shows that there were seventeen sightings of the nine obligate-focal species in restored sites, whereas there were ten sightings of the birds within the category of unrestored sites correcting for these definitions issues. This table developed a cumulative score for obligate bird sightings for the unrestored sites by removing the optimum wild-habitat site at Alverado Park from the unrestored category. The table also removed the Rheem Creek site used in the study under the category of a restoration site because of its weak link to restoration criteria. The resulting corrected comparison suggests that there may be significant differences between the restored and unrestored sites that enabled the restored sites to attract obligate-focal species. Research using a greater number of site comparisons and tighter definitions of the terms *restoration* and *unrestored* could provide more valid statistical comparisons.

Because of the number of confounding variables affecting this particular research design, the authors recommended that future studies address the pre- and post-project differences of bird populations at the restoration site. Now that we are more aware of the potential value of even neighborhood-scale restoration projects for bird refugia and the advantages of using pre- and postrestoration methods, we will now be prepared to conduct more pre-project bird surveys before we start construction of planned restoration projects.

The regional context for this study includes research conducted at the Coyote Creek Riparian Station near the City of San Jose on south San Francisco Bay for a number of years on riparian species. Bird species were trapped in nets, banded, and recaptured. Using the yellow warbler as an indicator species for riparian bird populations, the study found that a small restoration project along Coyote Creek

TABLE 4.3.

Focal and obligate species at restored and unrestored sites

Focal-Obligate Species	Number at Restored Sites	Number at Unrestored Sites
Black-headed grosbeak	1	0
Song sparrow	3	3
Swainson's thrush	0	1
Warbling vireo	3	2
Willow flycatcher	1	0
Wilson's warbler	4	1
Yellow warbler	4	2
Orange-crowned warbler	1	0
Common yellowthroat	0	1
Total	17	10

brought in more warblers that were using the installed willow and cottonwood habitat than showed up in the control conditions. Most importantly, the research indicated that this small urban stream restoration project functioned to provide critical migratory resting and feeding habitat in which the weight of the birds increased and enabled recovery from migration stress (Okamoto 1997). In an area such as San Francisco Bay, with its fragmented habitats, scientists are calling attention to the unique role that riparian habitats will provide in climate change adaptation because riparian plant species are particularly resilient to seasonal and annual variations in precipitation and temperatures (Seavy et al. 2009; Weiss, Shafer, and Branciforte 2010).

FISH AND WILDLIFE

Fish habitat restoration was not a primary objective of these neighborhood-scale projects. Attention to that need focused on the nearby larger regional-scale projects occurring on Sausal, San Leandro, Wildcat, Pinole, and Codornices Creeks as well as on Napa River and Napa Creek, which are not addressed in this book. The Strawberry Creek native stickleback came into view as a victim of chloramine poisoning by accidental discharges from potable water supply lines; likewise, the Pacific chorus frog was killed off for a period in the Baxter Creek watershed for the same reason. The Alhambra Creek beaver became important restorationists for the steelhead in this watershed, an accomplishment probably not to be replicated by human interventions. The other less charismatic but certainly most studied aquatic species in urban aquatic research projects are the benthic insects, which form an important part of the food chain for other aquatic and terrestrial species in stream corridors.

The new appearance of aquatic mammals such as beaver, otter, mink, and muskrat in densely populated centers has generated an awareness of the increase of wildlife living among urban dwellers. We now need to learn how to best support this phenomenon by developing strategies to reduce conflicts between urban human needs and wildlife habitat needs. One of our most constrained creek restoration sites on Poinsett Avenue now has two sightings of mountain lion. Unlike the beavers, which move into downtowns and urban streams to stay, the increasing sightings of mountain lions in very urban settings do not represent a reoccupation of urban spaces; they are temporary strays. In this case, public education is the critical component to human–wildlife cohabitation so that the public has a realistic perspective about the relative lack of threat from a species that is careful to avoid humans. Public education also needs to reduce the factors that can increase conflicts between wildlife and humans (Grand Canyon Trust 2013; Pollock et al. 2015).

Just as citizens' and nonprofit organizations have elevated the discovery of salmonid populations previously unknown, they have also initiated efforts to identify and publicize discovered urban beaver colonies. Citizens have increased awareness of the restoration functions that beavers can provide to aquatic environments and have taken the lead in changing how government agencies view the role of beavers in potentially returning historic ecological systems. The beaver protection group Worth a Dam, which was organized as a result of the Alhambra Creek case, set in motion a statewide California Beaver Working Group in 2012. This group included participants from nonprofit wildlife organizations specializing in human-wildlife conflicts as well as state and federal resource agencies staff. The formation of the working group was cohosted by the Occidental Arts and Ecology Center in Sonoma County and Worth a Dam, both of which were starting to track some of the salmonid restoration efforts that were integrating beaver habitat and collecting historical documentation that beavers were occupying California in the 1800s and much earlier.

Because the North American beaver was not considered to be native to the watersheds of coastal California except for a few limited areas in the far north of California, the management of the species for ecological enhancement of waterways was limited. The California Department of Fish and Wildlife accepted reports from the 1930s and 1940s, prepared by well-respected wildlife biologists (Grinnell, Dixon, and Linsdale 1937; Tappe 1942), who concluded that beavers were not native to most of California and that any beavers encountered represented nonnative survivors of twentieth-century relocations of populations into the state. This nonnative classification for beaver means that the state wildlife management agency does not recognize the species as one it should manage and protect as a native species and therefore that its only management activity is to issue depredation permits for the removal of the animal as a pest species.

The California Beaver Working Group agreed to a two-pronged work plan in which its members are elevating the awareness within the stream restoration community about the opportunities beavers can provide to enhance salmonid populations and other aquatic species and collecting reports that identify beaver historically occupying stream environments in California. The education efforts have focused on disseminating the research of Michael Pollock of the Northwest Fisheries Science Center of the National Oceanic and Atmospheric Administration (NOAA), who has collaborated with fluvial geomorphologists and fisheries biologists to measure the habitat functions that beaver ponds provide salmonids in the Pacific Northwest. The role of beavers in the "perennialization" of streams that would otherwise run dry for parts of the year, creation of deep-water habitat for fish, and discovery that the beaver ponds were typically not migration barriers have elevated the importance and potential of using beavers as a means of passively achieving restoration of salmonid habitat. The working group has also stressed the beaver's unique function as a "keystone" species that creates habitat for other wetland species as well as serving other desired functions such as improving water quality (Pollock, Press, and Beechie 2004; Pollock, Beechie, and Jordan 2007; Lanman et al. 2013; Lazar et al. 2015).

In 2013, a team of seven—nonprofit organizations, volunteer researchers, and NOAA staff—completed a study in which historic evidence of beaver presence in California establishes it as native species. The historic evidence includes physical evidence of beavers found in museum collections; archaeological information; first-person accounts by scientists, trappers, and rangers; and ethnographic information, including place names and newspaper accounts (Lanman et al. 2013). The next steps in this strategy are to present the information to the California Department of Fish and Game Commission and petition for the beaver to be managed as a native species.

BENTHIC BIOASSESSMENTS

The benthic insect assessments at the Baxter Creek Poinsett Avenue site described in chapter 3 gave us an opportunity to evaluate the strengths and weaknesses of this assessment as an indicator of restoration success. In this case, bioassessment indexes went up and down over a period of nine years. The basic research conclusions, frequently referenced in other urban benthic assessment literature, were that there are limits to the ability of urban stream restoration projects to achieve biological and water quality improvements and that rapid assessments can be used to determine a biological recovery potential in urban streams within two years. These studies acknowledge that there are typically numerous confounding variables that can limit the ability of a restoration project in an urban setting to achieve biological improvements in addition to biological reactions to the flashier urban runoff conditions (Purcell, Friedrich, and Resh 2002; Purcell 2004; Chin et al. 2009). Using that as a starting point, shouldn't our watershed assessments focus on identifying the potential sources of the limitations or stressors that are preventing biological recovery in a watershed? Were the nine years of bioassessment study results on Baxter Creek indicating limitations for biological recovery the results of the limited size of the restoration project? Was there something about the restoration project itself that was not designed well enough to re-create water quality functions of the creek? Are there land uses in the watershed that could be suspects for polluted runoff or point source pollution that continually suppresses benthic populations (Purcell 2004)? Is it really true that, given the number of variables in play and options for watershed improvement, the end point of benthic research is two years to establish the ultimate conditions achievable from restoration as concluded from this report? These questions are posed here to stress the need to put greater emphasis on presenting bioassessment results within a larger, more complete watershed context. In the Baxter Creek watershed, for example, a large upstream golf course is a place to look for causes of habitat limiters because of the likelihood of stormwater runoff containing pesticides and nutrients. The Baxter Creek biological assessment within a watershed context would also record the beginning of the new East Bay water treatment system in which chlorine drinking water treatment was substituted with chloramines, a much stronger treatment chemical with a long residual life. As described in chapter 3, this information first surfaces in the form of frog kills noticed by neighbors who could correlate the timing with unplanned discharges of treated drinking-water supplies flooding down the street and entering the creek. Chloramine, a powerful, long-lasting disinfectant that is deadly to amphibians and fish, was, according to East Bay Municipal Utility District and water board records, likely beginning to enter Baxter Creek as early as 1999 and continued to at least 2005. During the same period as the research projects were being conducted, neighbors reported a strong odor of the now-banned insecticide diazinon that a neighborhood resident was most likely using for ant control.

What can the Friends of Baxter Creek learn to help regulatory agencies address the steps needed to improve the biological life of the creek, and what can we learn about how to design a better restoration project from this research? Obviously, these issues were not involved in the research objective of the studies on Baxter Creek, but these questions are meant to impart the perspective of a practitioner who hopes that they frame future urban stream research assessments.

An important distinction therefore needs to be made in the assessment of urban stream restoration projects and biological health. Some studies suggest that given the challenges of urban environments, it is not possible to re-create stream health and biological diversity with riparian and stream channel restoration projects. Returning the structure, processes, dynamics, and some functions to urban creeks is a start to the repair of the ecosystem environment even if the stream functions are impacted by pollutants. Investigating the causes of the pollutants can then steer us toward addressing these factors as part of our restoration efforts as well.

Project Installation

Revegetation Strategies

The schools of riparian recovery include landscape design, soil bioengineering and functions recovery, large-scale horticultural, and passive restoration. These schools represent a different perspective on the objectives of the restoration projects as well as the methods of restoration. The methods range from traditional landscaping and landscape design practices using nursery container stock to the very functional soil bioengineering field that uses bundled cuttings and harvested tree limbs for stabilizing damaged or recently excavated environments, the use of farm equipment to plant out large numbers of rooted stock in large areas of newly acquired floodplain, and the return of more historic hydrologic conditions to floodplains to encourage self-recovery.

Chapter 2 discussed the tension between a traditional practice of landscape architecture, or the landscape design school, in which a primary purpose of the riparian planting is to create an attractive landscape, employing a human-based motivation for the design and restoration. This school is contrasted to the newly evolving practice of advancing aquatic and terrestrial ecological functions and enabling a dynamic stream environment in an urban setting without the use of rock and concrete. I call this the soil bioengineering and functions recovery school because it promotes the concept that revegetation strategies can have desirable engineering functions for erosion control and planform stability of rivers as well as function as habitat. This strategy produces an unmanicured habitat environment, as opposed to a well-controlled, garden-style environment. Another school related to riparian restoration that appears in the literature is the horticulture-based large-scale floodplain restoration, which uses traditional farm machinery to plant out large numbers of plant stock specifically grown by the restoration project for use on the site. A few species are matched with flood elevations and soil types to cover newly graded floodplain areas (Griggs 2009; Hammond, Griggs, and Gilbert 2011). We have observed the ability of floodplain riparian vegetation to re-establish on its own if measures are taken to return a more historic hydrology to a floodplain area.

Riparian restoration objectives can include increasing shade and reducing water temperatures, improving water quality through the capture of sediment and uptake of nutrients and pollutants, restoring input of organic matter and wood to streams, providing stream channel complexity for fish and aquatic life, creating terrestrial habitat, reducing erosion, and stabilizing stream banks. Restoration strategies can include forestry and grazing management, such as animal exclusion, riparian corridor protection, creation of riparian buffers, and active revegetation. For some practitioners of stream restoration, the ultimate objective is to recover wildlife habitat that only riparian corridors can offer, with this objective being a primary measurement of success in a project.

The practice of riparian or native plant restoration along streams shares a number of important similarities with the other disciplines involved with stream restoration. Passive as well as active restoration methods are applied. Historic data and reference sites as well as information on land use changes and hydrology and landscape classification systems are used to inform restoration design. As in the fields of geomorphology and fisheries ecology, the re-creation of functioning landscapes has to both recognize the value of historic landscape information and the existing context of landscape changes. The field also grapples with the relative values of small- versus large-scale projects.

Restoration guidance for riparian areas generally advises using mosaics of plant structure and density to provide a range of nesting, foraging, and cover for wildlife and birds. The planting structure usually includes a cover of high-canopy trees, a middle shrub layer, and layers of ground cover. Internal thickly planted forest refugia can provide some of the best habitat, and the outside edges of these dense areas can provide more layers, openings, and patchiness to maximize habitat values. Biologists report that the cottonwood–willow gallery forests of the Southwest support the highest concentrations of native nesting birds for North America (FISRWG 1998). Research has also established the importance of floodplain and riparian environments for fish to increase growth and survival rates. For example, the Sacramento spittail spawns on flooded floodplains and attaches it eggs to submerged vegetation (Sommer 2001; Feyrer et al. 2004).

It is interesting to note the similar functions riparian areas provide for both fish and birds and the overlap of restoration design needs and features. The statement that designing for bird habitat is designing for fish habitat, however, is being challenged by fisheries biologists. Obviously, birds and fish will use different parts of the riparian zone for cover from predators, and in fact, some birds will predate on fish. Birds like thickets to hide from their predators, and these hiding places are in thick shrub layers and canopy zones. In contrast, fish use low-hanging bank vegetation over the surface of the water to escape from wading birds like egrets and herons that hope to make the fish a meal (Ferguson 2013).

Riparian vegetation has been viewed historically as the "enemy" of river engineers, particularly when addressing single-purpose flood control, levee, or stream bank stabilization projects and the stability of creek-side structures. This view is that riparian vegetation compromises or competes against engineering objectives by reducing channel capacities for flood conveyance, causing backwatering of flood flows, trapping debris, causing bank failures, displacing rock riprap, making levees hard to inspect, compromising structural integrity of levees and berms by creating the hydraulic movement of water through levees referred to as piping, and harboring wildlife that can burrow in levee structures. Research hydraulic engineers as early as the 1980s debunked much of this engineering theory about the structural harm that vegetation may cause to levees (Riley 1981; Shields 1991). In fact, the research began to move in the opposite direction, indicating that not only is the vegetation not a threat to berms or levees but that the vegetation can add to levee strengths and stabilities and thereby both prevent catastrophic levee failures and complement engineering objectives of river projects (Shields and Gray 1992; Fischenich and Copeland 2001; National Resources Conservation

Service 2007a; Riparian Habitat Joint Venture 2007; Chen et al. 2009; US Army Corps of Engineers 2011).

With the reintroduction of soil bioengineering, also called biotechnical slope control, into stream engineering practices from its initial debut in the 1930s, river management has come full circle on this issue of the positive engineering functions of plants along streams and rivers. Riparian plants bundled and installed using the principles of soil bioengineering are now used as structural engineering components of stream project designs. The uses of plant material for engineering functions has entered quantitative analyses in which engineers calculate the stabilizing performance of plant systems. Engineers have established that the practice of plant-based soil bioengineering can equal or exceed the tensile strengths of the traditional engineering medium of concrete to stabilize a stream. Donald Gray, chair of the engineering department at the University of Michigan, while partnering with a landscape architect from Canada, Robbin Sotir, pioneered the merging of the field of mechanical engineering with plant physiology and ecology to introduce the role of plant material as an important structural function for stream projects (Gray and Sotir 1996). Design tables are now published by the US Army Corps of Engineers (Fischenich 2001) and the Natural Resources Conservation Service (2007b), which list the capability of soil bioengineering plant systems to resist flow velocities in feet per second (permissible velocities) and permissible shear stresses measured in pounds per square feet. The plant systems are listed along with conventional rock riprap stabilization systems so that their parallel performance can be compared against the more conventionally used erosion control systems such as rock riprap, which interfere with natural stream processes.

The most prevailing and consistent lessons across all the cases with the most significant impact on the project results came from evaluating the revegetation strategies used. The practice of revegetation began with creating long lists of potential East Bay native riparian restoration plants. Each of our creek restoration design reports contained a list of at least thirty plants that were "appropriate" to East Bay riparian corridors. Plants selected for the revegetation projects were often influenced most by what were favorite riparian natives (or, in the case of Baxter Creek at Poinsett Avenue, favorite drought-tolerant plants) of the designer or what was available to order from a commercial nursery. By 1995, the East Bay had a nonprofit plant nursery established by the California Native Plant Society that could grow advance-order riparian species not typically carried by commercial nurseries. This arrangement made it possible to grow the native blackberry groundcover at Blackberry Creek. Soil bioengineering systems were planted, and at the same time or shortly after, a variety of container species were added to the riparian corridor. Monitoring indicated that after five years, many of the container plants were not survivors. It eventually became clear that just because a plant was a California native riparian did not mean that it was going to survive in a riparian planting project.

Both neighborhood-scale as well as larger and more complex regional projects taught us the same lessons. For this reason, the section below provides a detailed overview of the cumulative lessons on revegetation from East Bay restoration sites beyond those described in this book. The primary lesson is that rather than working off appropriate native plants lists, we should have given more attention to understanding what species grew in these sites historically and then finding reference sites that would better inform plant selection for a particular location. The second important lesson was that placing out a diverse pallet of shrubs and other understory plants after construction was putting many plants under stresses they were not capable of withstanding. Understory plants did not survive unless they were fortunate enough to be under existing shade. We forgot to take the clue from their designation as "understory species." Even though most projects planted out about twenty species of plants, the typical survival after ten years or more was attributed to just a few pioneer species that were well suited to disturbed conditions and a few tree species. If we are to succeed with creating diversity in a plant restoration corridor, we need to wait three to five years for the pioneer species such as willow, cottonwood, and dogwood to succeed. They then create the microclimate required for understory species to become established.

Table 4.4 records the number of species planted in the East Bay restoration projects that survived over time during a two- to thirty-year period. I stored the as-built plants lists for all the projects in chapter 3 so as to record and compare survival through time. Survival numbers do not necessarily correspond with species that thrived, with table 4.4 recording a species even if there is only one representative plant left. Some species, such as ferns and many of the ground level herbaceous plants, reliably died out within two to three years. A consistent few species survived well over time, with willows (*Salix*), cottonwoods (*Populus*), alders (*Alnus*), oaks (*Quercus*), and maples (*Acer*) serving as the dependable structure for the riparian canopy. A few shrubs, such as dogwood (*Cornus*), coyote bush (*Baccharis*), toyon (*Heteromeles*), and wild rose (*Rosa*), were the consistent understory survivors. The native blackberry (*Rubus*) and ninebark (*Physocarpus*) performed well over time as ground covers.

To check the East Bay species survival list in Table 4.4 against a wider base of experience and environments in the Bay Area, I interviewed a number of practitioners involved in restoration revegetation projects from other locations of the bay. The appendix to this book lists the species that consistently survived over time and those that we placed on our risk list for long-term survival. The consistent list of survivors were eight species of trees, five shrubs, two types of ground cover, one vine, and one herbaceous plant, which matches well with both the numbers and

TABLE 4.4.

Restoration Project	Number of Species Planted		Number of Species Remaining
Strawberry Creek	**1983** (1)		**2012** (2)
Herbaceous	1		0
Ground cover	0		0
Shrubs	14		2
Trees	15		5
Vines	1		0
Glen Echo Creek	**1985** (3)		**2012** (4)
Shrubs	8		2
Trees	7		4
Vines	1		1
Ferns	1		0
Volunteers:			
Native trees			2
Nonnative trees			2
Native ground cover			1
Nonnative ground cover			1
Blackberry Creek	**1995–1996** (5)	**2000** (6)	**2012** (7)
Herbaceous	2	1	1
Groundcover	2	2	2
Shrubs	12	8	4
Trees	10	7	6
Vines	1	0	0
Ferns	2	1	0
Volunteers:			
Nonnative ground cover/shrubs	0	2	1
Nonnative trees	2	0	2
Baxter Creek at Poinsett Avenue	**1997** (8)		**2012** (9)
Herbaceous	3		0
Groundcover	0		0
Shrubs	11		6
Trees	5		4
Vines	2		2
Neighborhood-added trees	2		2
Nonnative shrubs	4		0
Nonnative trees	1		0
Volunteers:			
Nonnative ground cover	0		1
Baxter Creek at B. T. Anderson Park	**2000** (10)	**2010** (11)	
Herbaceous	8	0	
Groundcover	1	0	
Shrubs	11	5	
Trees	6	6	

TABLE 4.4. continued

Restoration Project	Number of Species Planted	Number of Species Remaining	
Supplemental Planting	**2010** (12)	**2011** (13)	
Herbaceous/rushes	10	7 (< 50% numbers survival); 3	
Ferns	2	2 (8–15% numbers survival)	
Shrubs	11	10 (14–100% numbers survival)	
Groundcover	2	2 (100% numbers survival)	
Vines	2	2	
		2014 (14)	
Herbaceous		6	
Rushes		0	
Groundcover		1	
Ferns		0	
Vines		0	
Shrubs		15	
Trees		6	
Village Creek (15)	**2000** (16)	**2009** (17)	**2012** (18)
Groundcover	1	1	1
Shrubs	6	4	3
Trees	6	5	5
Volunteers:			
Nonnative trees	1	1	2
Volunteers:			
Nonnative groundcover		2	2
Native groundcover			1
Vine native			1

Sources of plant counts: (1) Wolfe Mason Associates; (2) Waterways Restoration Institute (WRI); (3) Alameda County Flood Control District; (4) WRI; (5) Wolfe Mason Associates; (6) Junichi Imanishi, University of California, Berkeley; (7) WRI; (8) Brady Associates and WRI; (9) WRI; (10) Owens Viani; (11–13) City of Richmond (City of Richmond calculated percent survival of numbers of plants in 2011); (14–18) WRI.

types of species generally found in the East Bay creeks located in parks or open spaces.

Plant survival is more complicated than which species are selected for use in restoration projects, and in theory, most of the riparian species should be able to survive if planted with knowledge of soil conditions, rainfall, aspect, and exposure requirements. Maintenance is a big factor in survival rates, and there can be too much or too little. High mortality rates can be a factor of accidental damages from mowers, string weed trimmers, and maintenance staff or neighborhood volunteers who do not recognize which plants are intended to remain or the ecological objectives intended. The three common management problems are invasive weeds; browsing by deer, rats, and other wildlife; and damage by humans (often children) and their dogs. Other factors affecting survival rates are the quality of the stock acquired from nurseries and the season in which plants are installed.

Still other factors affecting outcomes have to do with the realities of restoration project implementation. Sites are often regraded and therefore offer an immediate open invitation for invasive weeds that otherwise would not occur. Irrigation systems have inconsistent maintenance or have serious management problems because of animal damage. Trampling of newly planted sites by the public and playing children can mean that a restoration site sustains early and profound damages. Even when restoration sites become shaded and relatively stable, the invasion of fennel, ivy, Himalayan blackberry, St. John's wort, and other nonnative species is assumed. Restoration may need to evolve to the acceptance that invasive nonnative species will be a permanent part of riparian corridors for which we do not have vigilant volunteer groups who remove them (Lennox et al. 2007). These factors also contribute to selection for the toughest pioneer species and the trees.

Doing a better job of studying reference sites provides the insight that we were trying to impose too much species diversity on some of the sites. I now have changed my perspective on how to measure revegetation success. It is not to make a count of numbers of species survival, but rather to evaluate whether the plants surviving function to provide multiple layers required for fish or bird habitat, assist sustaining the stream planform, and provide basic water quality benefits of temperature and nutrient control and support dissolved oxygen levels. The future challenge is to understand if specific riparian species provide unique functions for food or habitat that others do not. A good candidate would be the oak trees.

Maintenance

There have not been maintenance issues associated with excessive deposition or erosion in these neighborhood-scale projects. The design of relatively stable dimensions and slopes and stable spacing of a step pool channel have resulted in the desired low maintenance stability of the channels. Channel design cross sections assumed a large roughness value to support an unmaintained riparian corridor to avoid vegetation maintenance for channel capacity objectives. The maintenance issues that arose were associated with vegetation management. Village Creek and Blackberry Creek projects have received close to no maintenance. The use of erosion control fabric and densely placed posts and cuttings that accomplished quick, comprehensive cover were effective weed suppressors. The projects can remain relatively unmaintained after the early plant establishment period unless there is a desire to maintain view corridors across a site or avoid power lines.

Maintenance issues were most associated with accidental damage to plants from city maintenance crews or contractors and unsupervised plant removal by volunteer groups or unvetted maintenance activities that resulted in a disgruntled and surprised public. In a strange irony, a "friends of creek" organization took it upon itself to be in charge of maintaining creek environments and inflicted significant damage to habitats. Public complaints required regulatory agencies to respond to the environmental damage, which led to the understanding that public education at the time of the restoration project is critical to achieving restoration objectives. Project designers must address fears about homeless encampments and public safety with the public early in the design process.

The project sponsors did not anticipate the accidental or unintended vegetation removal or impacts caused by public or volunteer groups. Now that we know this issue may follow restoration projects, local government maintenance crews and volunteers who coordinate with local governments should be carefully trained. Optimally, federal and state natural resources and regulatory agencies should be holding trainings with local sponsors of projects and make revegetation and management criteria clear in project permits.

Comparing revegetation projects that have the primary objective to provide habitat against those that employ landscaping or gardening strategies clearly indicates that the gardening projects require many volunteer hours to pull weeds and prune to achieve the intended garden look. Adding a gardening with native plants component for aesthetics next to a restored riparian corridor can provide the public with a valued experience, and the garden does not have to detract from a functioning riparian ecosystem along the stream corridor. Unless there are gardening or park maintenance personnel whose job responsibility is to maintain this feature, however, it may be doomed to evolve to a weed patch overwhelmed by the more invasive exotic plants.

Project Costs: Design Build and Formal Bid Contracts

The case studies of neighborhood-scale projects indicate a substantial increase in costs, corrected for inflation as measured in cost per linear foot between 1983 and 2014. The numbers in table 4.5 tell a story of the evolution of urban stream restoration as a "cause" to a formally recognized business. A number of diverse factors converged, sending restoration into the realm of a relatively expensive industry. The factors include changes in labor practices and oversight, introduction of computer graphics, substantial changes in regulatory programs, and the evolving use of career restoration professionals (Prunuske 2014).

TABLE 4.5.

Neighborhood-scale project costs, with some comparisons to regional-scale project costs

Year Installed	Project	Linear Feet V/C[a]	Total Original Cost	Total Cost in 2014 Dollars	Cost per Linear Foot ($)
1983	Strawberry Creek	160/240	$60,000	$142,612	$594
1985	Glen Echo Creek (1)	225/247	$250,000	$550,037	$2,227
1995	Blackberry Creek	200/240	$155,000	$240,775	$1,003
1996	Baxter Creek, Poinsett	250/275	$25,000[b]	$37,721	$137
1999	Village Creek	700/940	$48,000	$72,424	$77
2000	Wildcat Creek flood project[c] (2)	5,000/6,500	$221,100	$303,962	$47
2000	Baxter Creek, Anderson Park	800/1,120	$150,000	$206,216	$184
2004	Codornices Creek Phase one[c,d] (3)	600/900	$1,805,000	$1,359,759	$1,510
2005	Baxter Creek, Gateway[d] (4)	750/950	$617,000	$747,906	$787
2005–2006	Wildcat Creek, Church Lane (5)	200/210	$226,700	$266,700	$1,270
2006–2007	Wildcat Creek, Rumrill Road	1,040/1,280	$304,000	$347,097	$271
2011	Napa River, Rutherford Reach Four[c,d] (6)	—/4,800	$5,100,000	$5,367,471	$1,118
2013	Wildcat Creek, Davis Park (7)	—/575	$1,396,800	$1,419,459	$,2468
2014	Napa Creek[c,d] (8)	—/3,540	$21,800,000	$21,800,000	$6,158

Note: Amounts shown are restricted to public outreach, planning, design, permitting, and construction for restoration components only. Trail and land costs are not included. Other costs listed below are shown in the dollar value of the year of construction. Cost per lineal foot is calculated by the meandering channel length (includes both stream banks).

(1) Costs include installation of gabions.

(2) Costs include $113,000 in US Army Corps planning and design costs. Land acquisition costs (not in table) were $3,688,000 (2008 dollars).

(3) Land acquisition: $2.8 million (not included in table).

(4) Trail, lighting, and streetscape improvements: $228,500; land acquisition costs: $447,000 (not in table).

(5) Costs do not include trail and senior center improvements.

(6) Costs do not include donated land.

(7) Costs do not include bridge and trail.

(8) Cost for land acquisition: $5.4 million (not included in table). Construction costs include removal of three bridges and construction of two pedestrian bridges.

[a] V = straight valley distance; C = meandering channel length.

[b] Culvert replacement without restoration: $38,795.

[c] Regional-scale projects.

[d] Bid contractors; all others design-build.

Early restoration projects in the 1980s and even to the late 1990s were generally conducted with little or moderate regulatory oversight. Regulatory agencies were thrilled that entities were stepping forward with the intent to improve the environment rather than make an impact. Restoration was a novel development, and the agencies wanted to be generous and lenient to these pioneers in the 1980s and 1990s. Soon after, however, the discussions and viewpoints on restoration became much more complicated when it became clear that some restoration projects were not well conceived or executed. The "practicing researchers" (researchers with experience in design) had gone into overdrive by the late 1990s, producing new and improved design guidance integrating restoration concepts as an alternative to conventional engineering. Regulatory agencies realized that they had a responsibility to make sure that the best available science was being applied (it was certainly a challenge to follow this moving target) and require greater attention to team-based, interdisciplinary, and collaborative design. This increased vigilance greatly increased planning times and costs.

The regulatory community was caught between the academic community, which was raising issues on the negative results of restoration projects and the lack of oversight on costs and results, and the practitioner, contracting community, which believed that the regulatory oversight was an unwelcome intrusion on its professional experience and judgment. This tension is still very much a part of the restoration scene. Some of the larger, regional-scale projects in the Bay Area illustrate the improvement in project quality when the tensions between the regulated and regulators give way to more friendly collaborative efforts. There is no doubt, however, that the increase in project costs reflect this development.

The evolution of computer graphic technology, great increase in costs of housing, skyrocketing increase in health care plans, rise in prevailing wages, and increase in energy costs also converged to create higher restoration project costs. Early restoration designers were content to use the hand-drawn plans of landscape architects for site layout, grading, and landscape plans. With the advent of the AutoCAD computer graphics program, which requires skilled, time-consuming work to create essentially the same product as drawn by hand, project design costs rose (Prunuske 2014). Although some believe that the AutoCAD method provides more professional-looking plans, the substance of the plans is not improved. Ultimately, contractors still believe that they need to use the new technology to compete. Even historic events such as the Hurricane Katrina disaster in New Orleans that required huge amounts of resources to rebuild and the increasing importance of Chinese industrialization and China's increasing need for materials added to the inflation of costs of materials and fuel in the first decade of the twenty-first century.

The other increase in costs is the evolution away from the design-build model of restoration projects to the bid and contract model. Most of the projects listed in table 4.5 are referred to as "design-build" projects in which nonprofit organizations became advocates for the projects, raised the funding for them from grants, set up the partnerships to achieve them, designed the projects, organized public outreach, and constructed the projects. A few for-profits, such as Prunuske-Chatham Inc., Restoration Design Group, and Balance Hydrologics, adopted the nonprofit model of project life administration, directing the projects through fundraising, public review, permitting, design, and construction. Prunuske-Chatham, for example, greatly prefers the efficiencies of the design-build model even as a for-profit company. The design-build model is cost efficient because the designer produces a detailed design plan but does not need to go through the process of drafting exhaustive specifications for a contractor unfamiliar with the site and restoration construction methods such as excavation of complex channel forms or soil bioengineering. This detailed specifications step is not necessary in the design-build model because the restoration designer also has the knowledge and capability to carry out his or her own design. The design-build projects listed here used small, family construction businesses as subcontractors with the equipment and expertise for demolition, excavation, grading, and hauling. The projects created new employment opportunities for these small businesses as well as for conservation corps.

The last two projects listed in table 4.5 represent the more current practice of the bid and contract system. This system significantly increases the cost of a project because the process of drafting specifications, producing bid packages, conducting a request for proposals, and qualifications reviews of competing contractors is expensive for both the contractors and the project sponsors. It also adds significantly to project planning times. The arguments for this system are that it opens up restoration projects to an open, competitive marketplace, enforces the use of prevailing wages for the labor hired, and prevents insider influences through political contacts from one contractor taking over an industry and squeezing out other qualified businesses. The system also is intended to protect public dollars by hiring the low bidders.

Six of the fourteen projects in table 4.5 represent the more current practice of the bid and contract system. Table 4.5 compares restoration costs only and not other project features such as land acquisition or trails. There are inherently some site conditions and project features that may make one project more expensive than another. Daylighting can cost more than other projects, although not necessarily. Of the top 50 percent highest-cost projects, five of the seven are contractor-bid projects. For comparison, both neighborhood-scale and larger regional-scale projects, with some involving the US Army Corps of Engineers, are listed. Some of the larger projects, such as the Napa River Rutherford Reach floodplain restoration project, have lower construction costs per linear foot than some of the smaller projects. The projects involving the US Army Corps of Engineers are exponentially higher in costs. The Napa Creek case is expensive for this reason, but it is also an example of using a low-bid contractor, which eventually created the added expenses of hiring better experienced contractors to help complete the work and correct for low-quality installations. A review of the average costs for reach-scale urban stream restoration projects using state grant records for the San Francisco Bay Area indicated an average cost of $800 to $1,200 per linear foot in 2014 dollars. A 2015 study evaluated stream restoration costs using information from forty-two projects, with a mean age of fifteen years and mean project length of approximately 2,300 feet representing more rural settings of Northern California. The study estimated that the design, planning, permitting, materials, and installation costs for the more complex channel-floodplain habitat restoration projects averaged about $289 to $963 per linear foot (Lewis et al. 2015).

How did this transition from design-build to bid-contractor happen, and what are the consequences? Legal issues on who may practice restoration and how much they must be paid surfaced as the restoration practice grew. This issue came to a head in California in 2004 when a union filed a grievance with the California Department of Water Resources Urban Stream Restoration Program. The union argued that the state-funded restoration project using student apprentices from a community college violated sections 1771 and 1774 of the California labor code, which requires the use of prevailing wages for all who work on public works projects. This conflict between those who wanted to provide community involvement, education, and training experiences in the restoration field with those who wanted to protect prevailing wages in this new field brought legal uncertainties to the restoration business. Many volunteer-dependent projects sponsored by government agencies and nonprofit organizations—such as coastal cleanups, invasive plant-removal projects, wetland revegetation projects, community-based tree-planting programs, and urban trash-removal projects—entered the realm of potential illegal activities if they met a definition of "public works " projects. The pushback came in the form of a campaign called "Use a volunteer, go to jail," organized mostly by nonprofit organizations, which educated the California legislature on the necessary ties between community involvement and public works projects. In 2004, legislation clarified the ability of local agencies to use conservation corps, volunteers, and students, but it was saddled with a so-called sunset clause that did not end the exemptions from the strict interpretation of the labor code until 2008. Repeated use of sunset clause in 2008, 2011, and 2015 on subsequent bills guarantees that this issue will continue to fester in California (Ames and Wellborn 2011). Bills to create a professional monopoly over who is certified to practice restoration have also been introduced in the state legislature by professional interest groups, such as foresters.

These issues created a legal conservatism in local governments that fear running afoul of contract and labor law. New requirements for publically funded projects include needing to offer to train union-sponsored apprentices if they are available. For now, the design-build model is mostly applied by private property owners or work funded by nonprofit land conservancies (Prunuske 2014).

Conclusions and Recommendations

For urban stream restoration to evolve to the next level, design practitioners and natural resources agencies that oversee project permitting will need to be more effective at explaining and promoting ecological restoration. We will all need to continue to evolve over time on how best to attain ecological restoration. Many current project funding programs do not address pre-project monitoring of fish, wildlife, and plant communities to compare post-project results against pre-existing conditions. Unless grant programs seek this information and request budgets for this activity as part of the grants, we will not make enough progress on linking stream restoration projects with in-stream and riparian habitat improvements. We will not be able to link physical habitat improvement projects with water chemistry and biological improvements until we begin to measure and evaluate the impacts of pesticides as limiting factors for ecological recovery. Ultimately, there must be stronger state and local government support roles for maintaining watershed partnerships to prioritize, plan, organize, and implement watershed improvements. Otherwise, it will be difficult for urban stream restoration to have a future.

Project Planning and Design

It is important for restoration designers in neighborhood settings to listen to the public and people's fears or issues about creating a new natural—and even "wild"—space in their environment. Equally important is that the designer must become an instructor, having a two-way conversation to balance restoration objectives and public needs for trail and road maintenance, power line protection, recreation, sports fields, and public safety. The public will not learn about the needs of native salmonids unless we explain the needs for temperature controls, shade, and complex channel environments that involve overhanging vegetation. They have no reason to know how different layers of vegetation will bring in birds that they and their children can enjoy watching and learning about. They have no way of knowing that the creeks provide important water quality services if well vegetated. They have no way of knowing that one of the best ways to stabilize streams is by using soil bioengineering systems, which has a wild and unkempt look that urban dwellers who are not used to experiencing natural environments may find unnerving. Landscape architects or planners who are not conversant in these aspects of stream restoration can solicit assistance from fish, bird, and water quality biologists who can be part of the design and communication teams.

These cases show that neighborhood-scale project designs are achievable by using relatively inexpensive applications of hydraulic geometry and reference site data and by being familiar with watershed conditions. Regional curves contribute important first draft designs that can be checked against regional hydrologic data and further confirmed with data on shear stresses and critical shear stresses. Simple use of Manning's equation or, in select cases, applying a HEC-RAS water surface model, can provide additional checks on design dimensions, discharges, and velocities if there is a clear flood risk reduction objective. In most of the neighborhood-scale projects, this type of modeling was not needed, but if it was used, it became a planning, not a design, tool. Project designs and construction budgets, minus major site structural elements such as lighting, benches, and trails, should try to achieve a restoration planning, design, and construction budget in the range of $800 to $1,200 per linear foot in 2014 dollars. Those who have the responsibility of reviewing projects for reach-level restoration should seek explanations for budgets that go over these ranges.

The use of prevailing wages for restoration work and the engagement of well-paid professionals is a positive development in the field of restoration. Prevailing wages are generally higher in urban areas than rural areas, so urban stream restoration projects reflect those wage increases. As the profession evolves, I hope it is recognized that there is not a conflict between community involvement and the use of conservation corps with creating good, prevailing-wage jobs and likewise that there is no conflict between paying prevailing wages and using the more cost-effective design-build model for the best efficient use of limited public funds.

Managing the Homeless and Fears of Habitat in Cities

Homeless people are best helped through local social service programs that specialize in the needs of the homeless and who have personnel trained to introduce the homeless to mental health services and housing programs. Clearing out riparian vegetation along streams to address public fears of the homeless is not a successful homeless management strategy, and local governments that find themselves spending scarce local resources on this activity will get better, cost-efficient results by using social service programs instead.

From my experience interacting with urban stream and river groups nationally, the public fears that can accompany creating habitat in cities appear to be centered in the western United States. Urban easterners and midwesterners, such as the public who live along Rock Creek in Washington, D.C., or the North Fork of the Chicago River in Chicago, appear to accept without consternation the presence of forests in their neighborhoods. An interesting research project would be to explore if fears of wild habitats in urban spaces can be correlated regionally. It is easy to speculate that because riparian corridors are less lush and tend to be of limited width in many southwestern environments, people in these locations are more unfamiliar and fearful of riparian woodlands. One of the fears about these riparian corridors is that homeless people will occupy the stream corridors and create a criminal element in the neighborhood.

Homelessness is a ubiquitous issue, particularly in the moderate climates of the Far West. My experience traveling over most of the state of California visiting urban streams is that the homeless will occupy areas along streams with lots of vegetation or with absolutely no vegetation at all. I could find no correlation with healthy riparian corridors and homeless occupation. The homeless occupy open or public spaces and stream corridors with or without vegetation. We learned that managing vegetation was not a homeless management strategy at the first conference of the national Coalition to Restore Urban Waters Conference in San Francisco in 1993. A homeless contingent from the American River Parkway in Sacramento attended this conference because they considered themselves stewards of the river. These representatives of the homeless were amused at the efforts of public works departments to discourage and disperse homeless encampments by pruning back and or removing riparian vegetation. The pruning is a greatly desired enhancement for their encampments because it provides space under trees for fitting in tents, beds, seating, and storage of possessions.

Because managing urban riparian corridors must often also manage for the impacts of the homeless on trash, water quality, fish, and wildlife habitat, we need to find a way out of the unproductive cycle of vegetation removal, rounding up homeless encampments, and confiscating possessions only to have the homeless arrive back again to the same sites. Someone who observed this cyclical failure is Herman Garcia, who grew up fishing for steelhead in the Pajaro River watershed in central coastal California, which includes Monterey, Benito, and Santa Cruz Counties. In 2006, he partnered with the National Marine Fisheries Service to address the crisis of the steeply declining salmonid populations in the 1,800 miles of stream channels in this large watershed containing a mix of rural and urban land use. Garcia met with the people in the homeless encampments and negotiated agreements to exchange river cleanup work to clean up waste and garbage for a daily hot meal and drinking-water delivery every two weeks along with some basic personal provisions. The homeless stewards are also involved in rescuing stranded steelhead due to occasional dewatering of sections of Uvas Creek, a tributary to the Pajaro. Their rescue efforts grew from saving a few hundred juveniles to tens of thousands of juveniles (Ambrose 2014; Garcia 2014). Garcia's group, Coastal Habitat Education and Environmental Restoration, or CHEER, advocates that any habitat restoration efforts first be preceded by this type of constructive engagement of homeless populations. The group's motto for successful engagement with the homeless is "show compassion, tolerance and then provide a purpose" for their lives.

We cannot disregard the sometimes immense impacts of the homeless on the quality of both rural and urban streams. In 2014, the Friends of Los Gatos Creek reported to the San Francisco Bay Area environmental regulatory agencies the horrible impacts that the homeless were making to this watershed. Not only were the homeless contributing untreated waste to the creek and living in trash-filled encampments, but they were seining the threatened steelhead out of the creek to eat. In 2014, these fish were under great threat from one of the worst droughts in California history, and this action is a heinous lack of disregard to the efforts of all the rest of us to restore the creek and fish populations. A coordinated effort involving state and local agencies and the friends of the creek group had to be organized to clean up the site and provide new living arrangements for the homeless.

Project Assessments

BIRD POPULATION HABITAT DEVELOPMENT AND RESEARCH

The authors of the study on the relationship of focal bird populations with restoration projects recommend that restorationists design their projects to encourage the fastest growth of habitat layers as possible (Demers and Scullen 2010). They encourage implementing large-scale restoration projects if possible. After the authors review the overlapping confounding variables in the study of restoration sites and bird populations, they also wisely recommend that future study designs emphasize pre- and postrestoration population surveys as opposed to trying to compare different restoration sites with unrestored sites. One problem with bird population study designs comparing different locations is they may not take into account that birds moving through an urban area will occupy degraded environments because that is all that is available in the subregion in which they are located. The inventories of before-and-after restoration conditions help develop better comparisons as a result. In fact, there is a perfect opportunity before us in that the study described here inventories bird populations at Wildcat Creek on Davis Park as an unrestored site in 2010, and in 2013, the site had the benefit of a newly installed restoration project. This site now has the beginnings of a post-project bird inventory. Bird population research has been clear that adding riparian corridors increases bird habitat. Research questions should start to address the functions that these bird habitats provide and possibilities for increasing these functions. Even moderately skilled volunteer birders located in the neighborhoods where these projects are located could observe which birds are present and how are they using the site. Are they foraging for insects? Are they foraging for berries or other food available in the corridor? Are birds nesting? If there are birds nesting, do they succeed with hatchlings and fledged young, and are there signs of loss through predation?

As researchers conduct more urban bird studies, we may want to conduct strategic assessments in which it is a priority to monitor for selected obligate and focal riparian species rather than conduct more time-consuming complete bird lists. According to the researchers involved in the East Bay restoration sites study, density of birds maybe the least reliable measure for assessing bird population response to restoration because density does not necessarily correlate with habitat quality. In fact, habitats of high quality have been shown to have low abundance, as discovered at the Coyote Creek Research Station near the City of San Jose. In this location, it was found that dominant individual birds can exclude other birds from a site (Okamoto 1997).

WATER QUALITY MONITORING

The next generation of water quality monitoring needs to focus scarce resources in a strategic manner to identify the toxic substances that are preventing biological recovery. Likewise, the EPA needs to be evaluating pest-control products for more than potential impacts to human health and should include adequate evaluations of potential impacts to aquatic life.

Given the chronic nature of pesticide use and other poisons affecting urban waterways, it is difficult to draw conclusions from bioassessment monitoring about the causes of chronic low- to moderate-quality benthic assessment indexes. The bioassessments have sent a strong message that biological integrity is compromised, and we need to move on to the question of asking why. Bioassessments have been widely used as indicators of relative water quality, but professional biologists warn that it is actually expensive to do these assessments with high levels of accuracy. We should use bioassessments more strategically to find where the highest benthic scores are located and identify if there are common conditions at these sites (Lunde 2014; O'Hara 2014).

The concept that many of our drinking-water treatment systems are toxic to aquatic organisms—fish, frogs, and amphibians—is new. If water agencies increasingly use more residual chemicals to treat water, we will need to consider that there may be significant impacts to endangered and threatened species when unintended discharges of drinking water enter waterways. The chloramine discharges described in the Baxter Creek case are particularly significant because they originate in the unstable top of the watershed, and as the discharges flow downstream through storm drains, the poison affects a large part of the stream system flowing to the bay. If the discharges occur during low flows in particular, the water can remained pooled in the stream behind culverts and other grade breaks for long periods of time. Upstream insects cannot flow down to downstream areas to re-establish because they may have also been affected. Depending on when an impact occurs, it may take up to a year for the recovery of the larval insects (Lunde 2014).

We will not succeed in addressing water pollution impacts to aquatic life without a focus on pesticide registration and use. The Baxter Creek case is an example of a project area subject to diazinon and chlorpyrifos of the organophosphate class of pesticides. The regulatory process eventually prohibited the use of these pesticides, but some people still use up the supplies on hand. After toxicity sampling in thirty-six urban streams, the San Francisco Bay Regional Water Quality Control Board found toxicity caused by the presence of diazinon and chlorpyrifos in all of them. The water board therefore assumed that these chemicals were ubiquitous in most streams and adopted a region-wide total daily maximum load standard for Bay Area streams to try to enforce the phase-out of these chemicals. The good news, according to a regional water board report, is that over time, the phasing out of these products in the marketplace does record results in the removal of the chemical presence in the streams. A 2007 regional monitoring report found that toxicity due to diazinon had decreased in urban creeks and that fish larvae survival rates were affected in only one sample (San Francisco Bay Regional Water Quality Control Board 2007). The bad news is that as soon as a pesticide product is taken off the market, another product is registered for use to replace it, which may be equally or more impactful than the product it replaced. Figure 4.11 was developed by the California Stormwater Quality Association to illustrate what it calls the never-ending "pesticide treadmill" of one environmentally damaging pesticide being replaced with another. In this case, the new products are pyrethoid pesticides, which may be worse that the pesticides they are replacing because of the very high toxicity levels to aquatic organisms and their long half-lives (O'Hara 2014).

A contributing cause of the chronic toxicity issues in urban creeks is that the EPA certification of these pesticides has largely focused on the potential human health consequences of the pesticide use rather than also looking at the potential impacts to aquatic life. Toxicity studies are not common as a means of identifying aquatic toxicity issues in streams and other aquatic environments because they are more expensive than the type of common background water quality monitoring currently conducted in most states. At this time, state water quality staff believe that pyrethoids are a likely candidate as the limiting factor of aquatic life in this region's streams, a condition that could be widespread. The expense of toxicity monitoring can be managed, however, by developing basic information on the likely stressors in a watershed, by developing reasonable hypothesis for causes, and then by strategically focusing the monitoring on the basic inventory. Monitoring sites and the timing of monitoring can be strategically planned (O'Hara 2014). By August 2014, as a result of a lawsuit by the Northwest Center for Alternatives to Pesticides and others in a case before the US District Court in Washington State, the EPA re-established creek-side no-spray buffer zones to protect endangered or threatened salmon and steelhead in California, Oregon, and Washington. The no-spray zones are imposed for carbaryl, chlorpyrifos, diazinon, malathion, and methomyl.

What is the best response to this situation? Public education and awareness are important because an informed public can affect government regulation and enforcement as well as allocation of resources to the problem. The use of "green

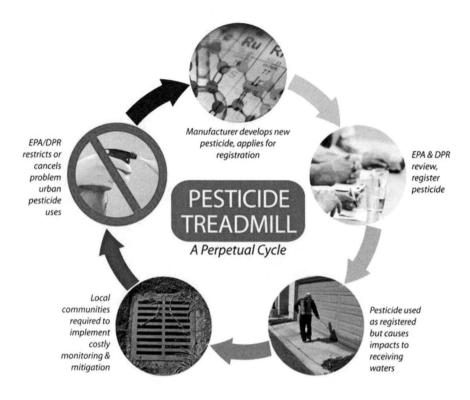

Figure 4.11 Until the "pesticide treadmill" is stopped, the biological integrity of streams will remain degraded. *Credit: California Stormwater Quality Association.*

stormwater" systems that infiltrate and remove some pollutants may contribute to lower toxicities in creeks because these pollutants are often attached to sediments that can settle out in stormwater catchments. In the East Bay, most green stormwater projects are being located in the downstream portions of watersheds to improve the water quality of stormwater before it enters the bay. The locations of these systems do not help address the creek channels upstream, where the discharges of chloramines or pesticides originate. Therefore, stormwater treatment locations need to be strategically located to assist upstream and downstream locations.

Functional Riparian Restoration

All the scientific disciplines converge on emphasizing the value of returning functioning riparian vegetation to stream corridors. The fluvial geomorphologist Richard Hey states that of all the parameters he could manipulate in his flume studies, vegetation performed one of the most profound influences on channel stability, and tree-lined stream banks are the most resilient to floods and land use changes

(Hey 1999). Fish and bird biologists recommend the layered riparian corridors for functioning wildlife habitat, and water quality biologists emphasize the important functions of the riparian corridor to improve or protect water quality. An extensive body of research correlates water quality with healthy riparian corridors (Dosskey et al. 2010).

Relevant research to this book's case locations has been performed by the San Francisco Bay Regional Water Quality Control Board. The board used an intensive monitoring effort in nine to twelve watersheds representing different regions and land uses. The study indicated that there is a significant water quality difference between stream corridors that are well vegetated with riparian vegetation and those that are not:

> Region wide there were significant correlations between riparian habitat conditions and measurements of temperature and dissolved oxygen. Sites at which channels were less modified where riparian and stream bank vegetation was more extensive and canopy cover was greatest had significantly lower stream temperatures and significantly higher dissolved oxygen concentrations. (San Francisco Bay Regional Water Quality Control Board 2007)

We can conclude that one important road to re-creating biological diversity is to engage the community to heighten awareness about pollutants. The other road to recovery is to replant the stream corridors with extensive vegetation and canopy cover.

The Keystone to Watershed Restoration: Citizen Involvement

One of the best investments that government agencies can make to further their mission to manage urban watersheds is to financially support citizens organizations that monitor, restore, provide public education about watersheds, and create training and jobs for the people living in the watersheds. In every case in this book, it was citizen involvement that initiated the restoration projects or intervened to turn conventional flood control projects into multi-objective projects. By the early 2000s, local governments that began to view urban stream restoration as a mainstream activity initiated projects. Project initiation and sponsorship need to be viewed as a complex unfolding of efforts over time, as illustrated by the Baxter Creek cases. Melanie Mintz, director of community development for El Cerrito, believes that her contribution to the Baxter Creek Gateway project was to "carry the baton the next step to help realize the ground work of Friends of Baxter Creek," which put years into organizing the community, raising awareness at city hall, writing grants for land acquisition, and pressing for restoration projects throughout the watershed (Mintz 2014).

Citizen organizations increase public awareness about the needs and opportunities for watershed environmental improvements and thereby provide the political base that mobilizes political representatives and ultimately provides the sources of funding. Networks and coalitions sponsor ordinances and laws to protect the resources and mobilize for pollution cleanup. A powerful tool in increasing public awareness was the development and publishing of urban creek and watershed maps for the urban Bay Area counties. Christopher Richards engaged the Oakland Museum in outreach efforts to friends of creek groups and local officials to piece together maps of underground and aboveground creek sections. These valuable maps provide basic classifications of stream conditions, show links to stormwater systems, and provide historic landscape information. As these maps were published, one by one, they literally put urban creeks on the map of public consciousness. The great irony is that although organized citizens are the base of the pyramid for environmental improvements, innovations, and policy improvements, it is the most fragile, vulnerable, and unfunded layer of the stakeholder process that must exist for the collaborations responsible for watershed improvements. Often, government programs view assisting capacity building for watershed organizations as something outside their purview. The stark reality, however, is that government programs for flood reduction, stormwater management, water quality improvement, and habitat enhancement will only be as strong as the organized public support and participation.

References

Ambrose, Jonathan. 2014. Fisheries biologist, National Marine Fisheries Service, National Oceanic and Atmospheric Administration, Santa Rosa, CA. Personal communication to author.

Ames, Laurel, and Michael Wellborn. 2011. "Statewide Volunteers, Threatened Again." California Watershed Network white paper.

Aparicio, Enric, Gerard Carmona-Catot, Peter Moyle, and Emili Garcia-Berthou. 2011. "Development and Evaluation of a Fish-Based Index to Assess Biological Integrity of Mediterranean Streams." *Aquatic Conservation* 21 (4):324–337.

Center for Watershed Protection. 2003. *Impacts of Impervious Cover on Aquatic Systems*. Ellicott City, MD: Center for Watershed Protection.

Chen, Z. Q. Richard, Kavvas, M. Levent, Hossein Bandeh, Elcin Tan, John Carlon, and Thomas Griggs. 2009. "Study of the Roughness Characteristics of Native Plant Species in California Floodplain Wetlands." Sacramento: California Department of Water Resources.

Chin, Anne. 2006. "Urban Transformation of River Landscapes in a Global Context." *Geomorphology* 79:460–487.

Chin, Anne, Alison Purcell, Jennifer Quan, and Vincent Resh. 2009. "Assessing Geomorphical and Ecological Responses in Restored Step-Pool Systems." Geological Society of America Special Paper 451:199–214.

Cluer B., and C. Thorne. 2014. "A Stream Evolution Model Integrating Habitat and Ecosystem Benefits." *River Research and Applications* 30 (2):135–154.

Collins, Laurel, and Roger Leventhal. 2013. "Regional Curves of Hydraulic Geometry for Wadeable Streams in Marin and Sonoma Counties." San Francisco Bay Area Data Summary Report, May 10. Prepared for San Francisco Estuary Partnership and US Environmental Protection Agency.

Contra Costa County. 2003. *Watershed Atlas*. Martinez, CA: Contra Costa County Community Development Department and Public Works Department.

Copeland, Ronald R., Dinah N. McComas, Colin R. Thorne, Philip J. Soar, Meg M. Jonas, and Jon B. Fripp. 2001. "Hydraulic Design of Stream Restoration Projects." Coastal and Hydraulics Laboratory Engineer Research and Development Center ERDC/CHL TR-01-28. Washington, DC: US Army Corps of Engineers.

Dearen, Jason. 2013. "River Otters Coming Back to the Urban Environment." *Contra Costa Times*, December 14.

Demers, Jill B., and Josh Scullen. 2010. "Bird Use of Urban Riparian Restoration Sites in Contra Costa County, San Francisco Bay Bird Observatory, California." Report prepared for the Urban Creeks Council, Berkeley, CA.

DeWeerdt, Sarah. 2014. "Cohabitation." *Conservation*, Winter, 33–39.

Dosskey, Michael, Philippe Vidon, Noel Gurwick, Craig Allan, Tim Duval, and Richard Lowrance. 2010. "The Role of Riparian Vegetation in Protecting and Improving Chemical Water Quality in Streams." *Journal of American Water Resources Association*, Paper no. JAWRA-09-0035-P:1–18.

Dunne, Thomas, and Luna Leopold. 1978. *Water in Environmental Planning*. San Francisco: Freeman.

Federal Interagency Stream Restoration Working Group (FISRWG). 1998. "Stream Corridor Restoration, Principles, Processes and Practices." Washington, DC: USDA Natural Resources Conservation Service.

Ferguson, Leslie. 2013. Water quality engineer and fish biologist, San Francisco Bay Regional Water Quality Control Board, Oakland, CA. Personal communication to author.

Feyrer, F., T. R Sommer, S. C. Zeug, G. O'Leary, and W. Harrell. 2004. "Fish Assemblages of Perennial Floodplain Ponds of the Sacramento River, California (USA), with Implications for the Conservation of Native Fishes." *Fisheries Management and Ecology* 11:335–344.

Fischenich, Craig. 2001. "Stability Thresholds for Stream Restoration Materials." Engineer Research and Development Center, Environmental Laboratory, EMRRP Technical Note Collection ERDC TN-EMRRP-SR-29. Vicksburg, MS: US Army Corps of Engineers.

Fischenich, J. Craig, and Ronald Copeland. 2001. "Environmental Considerations for Vegetation in Flood Control Channels." Engineer Research and Development Center ERDC TR-01-16. Vicksburg, MS: US Army Corps of Engineers.

Garcia, Herman. 2014. "The Homeless: Turning the Problem into the Solution." Presentation to the Bay Area Flood Protection Agencies Association and Bay Area Watershed Network Annual Conference, February 20, Oakland, CA.

Grand Canyon Trust. 2013. "Beaver: Best Management Practices, A Practical Guide to Living and Working with Beaver." Solutions to Life on the Colorado Plateau Grand Canyon Trust Utah Forest Program, Flagstaff, AZ.

Gray, D. H., and R. Sotir. 1996. *Biotechnical and Soil Bioengineering Slope Stabilization, A Practical Guide for Erosion Control.* New York: Wiley.

Griffin, John. 2015. Director of Urban Wildlife Program, Humane Society of the United States. Written communication to author, December 15.

Griggs, F. Thomas. 2009. *California Riparian Habitat Restoration Handbook*, 2nd ed. N.P.: California Riparian Habitat Joint Ventures.

Grinnell, J., J. S. Dixon, and J. M. Linsdale. 1937. *Fur-Bearing Mammals of California: Their Natural History, Systematic Status and Relations to Man.* Berkeley: University of California Press.

Hammond, F., Thomas Griggs, and Meghan Gilbert. 2011. "Long-Term Monitoring of Horticultural and Ecological Performance of Riparian Restoration Plantings along the Sacramento River, California, USA." Report prepared by River Partners, Chico and Modesto, CA.

Hey, Richard. 1999. "A Geomorphic Approach to Flood Hazard Mitigation and Emergency Response." National Park Service and Trout Unlimited Workshop. August 3–5, 1999, Livingston Manor, NY.

Lanman C., K. Lundquist, H. Perryman, J. Asarian, B. Dolman, R. Lanman, and M. Pollock, 2013. "The Historic Range of Beaver (*Castor canadensis*) in Coastal California." *California Fish and Game* 99 (4):193–211.

Lazar, Julia, Kelly Addy, Arthur Gold, Peter Groffman, Richard McKinney, and Dorothy Kellogg. 2015. "Beaver Ponds: Resurgent Nitrogen Sinks for Rural Watersheds in the Northeastern United States." *Journal of Environmental Quality* 44:1684–1693.

Lee, Liang. 2012. "South Bay Regional Curve for Bankfull Discharge versus Drainage Area." Graph. Santa Clara Valley Water District, San Jose, CA.

Lennox, Michael, David Lewis, Kenneth Tate, John Harper, Stephanie Larson, and Randy Jackson. 2007. "Riparian Revegetation Evaluation on California's North Coast Ranches." Final Report prepared by University of California Cooperative Extension, County of Sonoma.

Leopold, L. B. 1994. *A View of the River.* Cambridge, MA: Harvard University Press.

Leopold, L. B., and M. G. Wolman. 1960. "River Meanders." *Geological Society of America Bulletin* 71:769–793.

Leopold, Luna B., M. Gordon Wolman, and John P. Miller. 1964. *Fluvial Processes in Geomorphology.* San Francisco: Freeman.

Lewis, D., M. Lennox, A. O'Green, J. Creque, V. Eviner, S. Larson, J. Harper, M. Doran, and K. Tate. 2015. "Creek Carbon, Mitigating Greenhouse Gas Emissions through Riparian Restoration." University of California Cooperative Extension in Marin County Novato, CA.

Lunde, Kevin. 2014. Environmental specialist, San Francisco Bay Regional Water Quality Control Board, Oakland, CA. Personal communication to author.

Miller, Sarah, and Dan Davis. 2002. "Identifying Relationships for Bankfull Discharge and Hydraulic Geometry at USGS Stream Gage Sites in the Catskill Mountains, New York." New York City Department of Environmental Protection Stream Management Program.

Mintz, Melanie. 2014. Director of Community Development, City of El Cerrito, CA. Interview.

National Research Council. 2002. *Riparian Areas, Functions and Strategies for Management*. Washington, DC: National Academy Press.

National Resources Conservation Service. 2007a. "Streambank Soil Bioengineering Technical Supplement 141." Part 654, *National Engineering Handbook*, Table TS141-4. Washington, DC: US Department of Agriculture.

———. 2007b. "Stream Restoration Design." Part 654, *National Engineering Handbook*, 210-VI-NEH. Washington, DC: US Department of Agriculture.

O'Hara, Janet. 2014. Water quality engineer, San Francisco Bay Water Quality Control Board, Oakland, CA. Personal communication to author.

Okamoto, Ariel. 1997. "Warblers Refuel at City Creek." *Estuary* 6 (6):1.

Pollock, M., T. Beechie, and C. Jordan. 2007. "Geomorphic Changes Upstream of Beaver Dams in Bridge Creek: An Incised Stream Channel in the Interior Columbia River Basin, Eastern, Oregon." *Earth Surface Processes and Landforms* 32:1174–1185.

Pollock, M., G. Press, and T. Beechie. 2004. "The Importance of Beaver Ponds to Coho Salmon Production in the Stillaguamish River Basin, Washington, U.S.A." *North American Journal of Fisheries Management* 24:749–760.

Pollock, M. M., G. Lewallen, K. Woodruff, C. E. Jordan, and J. M. Castro, eds. 2015. *The Beaver Restoration Guidebook: Working with Beaver to Restore Streams, Wetlands, and Floodplains*. Version 1.0. Portland, OR: US Fish and Wildlife Service.

Prunuske, Liza. 2014. Interview. Principal of Prunuske-Chatham Inc., Sebastapol, CA.

Purcell, Alison. 2004. "A Long-Term Post Project Evaluation of an Urban Stream Restoration Project, Baxter Creek, El Cerrito, California." Restoration of Rivers and Streams, University of California Multi-Campus Research Unit. Water Resources Center Archives, Berkeley, CA.

Purcell, Alison, Carla Friedrich, and Vincent Resh. 2002. "An Assessment for a Small Urban Stream Restoration Project in Northern California." *Restoration Ecology* 10 (4):685–694.

Purdy, Sarah, Peter Moyle, and Kenneth Tate. 2011. "Montane Meadows in the Sierra Nevada: Comparing Terrestrial and Aquatic Assessment Methods." Center for Watershed Sciences and Department of Plant Sciences, University of California, Davis; Springer Science+Business Media B.V. June 28.

Rantz, S. E. 1971. "Suggested Criteria for Hydrologic Design of Storm-Drainage Facilities in the San Francisco Bay Region, California." Open File Report 71-341. Menlo Park, CA: US Geological Survey.

Riley, A. L. 1981. *Observations and Analysis of Levee Maintenance Practices.* Sacramento: California Department of Water Resources.

Riparian Habitat Joint Venture. 2004. *The Riparian Bird Conservation Plan: A Strategy For Reversing the Decline of Riparian Associated Birds in California.* California Partners in Flight. http://www.prbo.org/calpif/pdfs/riparian.v-2.pdf.

———. 2007. *Riparian Habitat Conservation and Flood Management in California.* Conference Proceedings, California Partners in Flight, December.

Roesner, Larry A., and Brian P. Bledsoe. 2003. *Physical Effects of Wet Weather Flows on Aquatic Habitats: Present Knowledge and Research Needs.* London: IWA Publishing.

Roy, A. H., and W. D. Schuster. 2009. "Assessing Impervious Surface Connectivity and Applications for Watershed Management." *Journal of the American Water Resources Association* 45 (1):198–209.

San Mateo Countywide Water Pollution Prevention Program. 2013. *Geomorphic Study in San Pedro Creek.* Oakland, CA: EOA Inc.

Schumm, S. A., M. D. Harvey, and C. Watson. 1984. *Incised Channels: Morphology, Dynamics and Control.* Littleton, CO: Water Resources Publications.

Seavy, N. E., T. Gardali, G. H. Golet, F. T. Griggs, C. A. Howell, R. Kelsy, S. L. Small, J. H. Viers, and J. F. Weigand. 2009. "Why Climate Change Makes Riparian Restoration More Important than Ever." *Ecological Restoration* 27:330–338.

San Francisco Bay Regional Water Quality Control Board. 2007. "Water Quality Monitoring and Bioassessment in Nine San Francisco Bay Region Watersheds." Oakland, CA: San Francisco Bay Regional Water Quality Control Board.

Shields F. D., and Donald H. Gray. 1992. "Effects of Woody Vegetation on Sandy Levee Integrity." *Water Resources Bulletin* 28 (5):917–931.

Shields, F. Douglas. 1991. "Woody Vegetation and Riprap Stability along the Sacramento River Mile 84.5–119." *Water Resources Bulletin* 27 (3):527–536.

Shields, F. Douglas, Jr., Ronald Copeland, Peter Klingeman, Martin W. Doyle, and Andrew Simon. 2008. "Stream Restoration." In *Sedimentation Engineering Processes, Measurements, Modeling and Practice* (Manual of Practice No. 110), edited by Marcelo Garcia, chap. 9. Reston, VA: American Society of Civil Engineers.

Simon, A. 1989. "A Model of Channel Response in Distributed Alluvial Channels." *Earth Surface Processes and Landforms* 14 (1):11–26.

Skidmore, P., C. Thorne, B. Cluer, G. Pess, T. Beechie, J. Castro, and C. Shea. 2009. *Science Base Tools for Evaluating Stream Engineering, Management, and Restoration Proposals.* NOAA Fisheries and US Fish and Wildlife Service.

Soar, Phillip, and Colin Thorne. 2001. "Channel Restoration Design for Meandering Rivers." Coastal and Hydraulics Laboratory Engineer Research and Development Center ERDC/CHL CR-01-1. Washington, DC: US Army Corps of Engineers.

Sommer, T. R., M. L. Nobriga, W. C. Harrell, W. J. Batham, and W. J. Kimmerer. 2001. "Floodplain Rearing of Juvenile Chinook Salmon: Evidence of Enhanced Growth and Survival." *Canadian Journal of Fisheries and Aquatic Sciences* 58 (2):325–333.

Sullivan, Joe. 2015. Fisheries program manager, East Bay Regional Park District, Oakland, CA. Personal communication to author.

Tappe, D. T. 1942. "The Status of Beavers in California." California Department of Natural Resources, *Game Bulletin* 3:4–41.

Thorne, C. R. 1999 "Bank Processes and Channel Evolution in the Incised Rivers of North-Central Mississippi." In *Incised River Channels*, edited by S. E. Darby and A. Simon, 97–122. Chichester, UK: Wiley.

Thorne, Colin R., Richard Hey, and Malcolm Newson. 1997. *Applied Fluvial Geomorphology for River Engineering and Management.* Chichester, UK: Wiley.

US Army Corps of Engineers. 2011. "Initial Research into the Effects of Woody Vegetation on Levees, Summary of Results and Conclusions." Engineer Research and Development Center. Vicksburg, MS: US Army Corps of Engineers.

Walsh, Christopher, Tim Fletcher, and Anthony Ladson. 2005. "Stream Restoration in Urban Catchments through Redesigning Stormwater Systems: Looking to the Catchment to Save the Streams." *Journal of the North American Benthological Society* 24 (3): 690–705.

Warner, Richard E., and Kathleen M. Hendrix, eds. 1984. *California Riparian Systems, Ecology, Conservation, and Productive Management.* Berkeley: University of California Press.

Weiss, S. B., N. Schafer, and R. Branciforte. 2010. "San Francisco Bay Area Upland Goals Draft." Riparian/Fish Focus Team Report, Bay Area Open Space Council, California Coastal Conservancy, US Fish and Wildlife Service, Oakland, CA.

Wise, Scott, Pete Alexander, and Matt Graul. 2007. "Fisheries Habitat Inventory and Assessment for Lower Wildcat Creek." Oakland, CA: East Bay Regional Park District for the Urban Creeks Council.

Regional San Francisco Bay Restoration Plant Survivors and Plants Associated with More Risk to Survival (Common and Latin Names)

Consistent Survivors

Near stream bank and floodplains

All willow species	*Salix* spp.
Cottonwoods	*Populus fremontii* (not Marin County)
Alders, red and white	*Alnus rhombifolia* and A. *oregana*
California blackberry	*Rubus vitifolius*
Ninebark	*Physocarpus capitus*
Dogwood	*Cornus stoloniferra* and C. *sericea*
Hawthorn	*Crataegus douglasii*
Creeping wild rye	*Leymus triticoides* (can compete in environments with difficult vulnerability to exotic invasives, particularly if planted as rooted container stock)

Midslope

Box elder	*Acer negundo*
Oregon ash	*Fraxinus latifolia*
Maple	*Acer macrophyllum*
California blackberry	*Rubus vitifolius*
California rose	*Rosa californica*
Dogwood	*Cornus sericea*
Wild grape	*Vitis californica*

Soil bioengineering

Willows	*Salix* spp.
Cottonwoods	*Populus* spp.
Dogwood	*Cornus sericea*

Top stream slope:

Buckeye	*Aesculus californica* (star performer in South Bay planted by seed)
Oaks	*Quecus agrifolia* and *Q. lobata*
Maple	*Acer macrophyllum*
Coffeeberry	*Rhamnus californica* (not listed as riparian species by Faber and Holland)
Coyote brush	*Baccharis piluaris*
Cottonwoods	*Populus* spp.

South Bay species

High slope:

California sycamore	*Platanus californica*

Low and midslope:

Clematis	*Clematis ligusticifloia* (with irrigation and a blue elderberry to climb)
Mugwort	*Artemisia douglasiana*
Blue elderberry	*Sambucus cerulea* (exceptional performer)
Mule fat	*Baccharis viminea* (in dry washes and inland areas, cuttings in low bank)

Second-Phase Understory with Good Survival after Some Canopy Cover

Lower stream bank

Thimbleberry	*Rubus pariflorus*
Douglas iris	*Iris douglasiana*

Midbank to high bank

Snowberry (a mixed record)	*Symphoricarpus* spp.
Cow parsnip	*Heracleum lanatum*
Spicebush	*Calycanthus occidentalis* (some shade, some sun)
Blue elderberry	*Sambucus cerulean*

High stream bank

Wax myrtle	*Myrica california* (more difficult to thrive inland)
Coffeeberry	*Rhamnus californica* (not listed as a riparian plant in some publications)
Toyon	*Heteromeles arbutifolia* (good results in East Bay, not as good in North Bay)

Low Success Rates, Risky

Chaparral species in general, including

Salvias	*Ceanothus* spp. (short lived but can survive for five-plus years before it is shaded out and water stressed; not good for long-term results in most situations)
Monkey flower	*Mimulus* (including the one riparian species *Guttatus*)
Buckwheats	*Erigonum* spp.
Ocean spray	*Holodiscus discolor*
Mountain mahogany	*Cercocarpus betuloides*
Manzanitas	*Arctostaphylos* spp.
All ferns	Many species have extremely high and quick mortality rates on newly planted sites; *Woodwardia fimbriata* has reports of doing well in full shade and roots in water
Dutchmen's pipevine	*Aristolochia californica*
Alum root	*Huchera micrantha*
Huckleberry	*Vaccinium ovatum*
Hazelnut	*Corylus cornuta*
Red-flowering currant	*Ribes sanguineum*
Golden currant	*Ribes aureum*
Bay	*Umbellularia californica*

Mixed records

California black walnut	*Juglans hindsii*
Honeysuckles	*Lonicera hispidula* and *L. involucrata*
Blue elderberry	*Sambucus cerulea*
Red elderberry	*Sambucus callicarpa*
Redwood	*Sequoia sempervirens*
Catalina cherry	*Prunus illicifolia*
Gooseberry	*Ribes speciosum*
Mule fat	*Baccharis viminea* (in North Bay)
Mugwort	*Artemisia douglasiana*

Sedges and rushes

Santa Barbara sedge	*Carex barbarae*
Pacific sedge	*Juncus effuses*
Common rush	*Juncus patens*
Bulrush	*Scirpus microcarpus* (will often come in as a volunteer and does not need to be planted)

Source: Compiled from information provided by Liza Prunuske, Steve Chatham, Harold Appleton, Mike Jensen, Maggie Young, and Joan Schwan. Prunuske Chatham, Inc.; Liz Lewis, Marin County; Keenan Foster, Sonoma County Water Agency; Jane Kelly and Carole Schemmerling, Urban Creeks Council; Lisa Graves, City of Richmond; Linda Spahr, Santa Clara Valley Water District; and A. L. Riley, San Francisco Bay Water Board.

Dr. Ann L. Riley is the executive director of the Waterways Restoration Institute (WRI) and is watershed and river restoration advisor for the San Francisco Bay Regional Water Quality Control Board, a California state agency. During her tenure with the nonprofit WRI, she has organized, planned, designed, constructed, and funded numerous stream restoration projects in California and throughout the United States. Her involvement in community-level nonprofit organizations and her work with local, state, and federal agencies in watershed planning, water quality, water conservation, hydrology, flood management, stream science, and restoration span several decades. Her long history of service includes public policy work for nonprofit organizations, the National Academy of Sciences, and the John Heinz Center for Science, Economics, and the Environment.

In 1982, she cofounded the Urban Creeks Council in California, and in 1993, she was instrumental in organizing the first conference of the Coalition to Restore Urban Waters, a national network of urban stream and river organizations. In 1984, she spearheaded a program under the auspices of the California Department of Water Resources that continues to provide grants supporting urban stream restoration. In both her private- and public-sector work, she has championed jobs and training for conservation and youth corps.

Dr. Riley's work in urban river restoration is recognized throughout the United States. She has garnered numerous awards over her career, including an American Rivers award in 1993 for her leadership in establishing a national urban river movement, the California Governors' Environmental and Economic Leadership award in 2003, and the Salmonid Restoration Federation Restorationist of the Year Award in 2004.

Her association with river scientist Luna Leopold reaches back to 1971 in Washington, D.C. Dr. Riley completed two graduate degrees under his direction at the University of California, Berkeley. She is an urban farmer at her home in Berkeley, California, where she raises chickens and bees, grows food, and home-brews mead and beer that win awards at state and county fairs.

Note: page numbers followed by b, f, or t refer to boxes, figures, or tables, respectively.